Remaking Housing Policy

Breaking the country-specific boundaries of traditional housing policy books, *Remaking Housing Policy* is the first introductory housing policy textbook designed to be used by students all around the world. Starting from first principles, readers are guided through the objectives behind government housing policy interventions, the tools and mechanisms deployed and the outcomes of the policy decisions.

A range of international case studies from Europe, Asia, Africa and the Americas illustrate the book's general principles and demonstrate how different regimes influence policy. The rise of the neo-classical discourse of market primacy in housing has left many countries with an inappropriate mix of state and market processes with major interventions that do not achieve what they were intended to do. *Remaking Housing Policy* goes back to basics to show what works and what doesn't and how policy can be improved for the future.

Remaking Housing Policy provides readers with a comprehensive introduction to the objectives and mechanisms of social housing. This innovative international textbook will be suitable for academics, housing students and those on related courses across geography, planning, property and urban studies.

David Clapham is a Professor of Planning within the Department of Real Estate and Planning, University of Reading, UK.

Remaking Housing Policy

An International Study

David Clapham

LONDON AND NEW YORK

First published 2019
by Routledge
2 Park Square, Milton Park, Abingdon, Oxon OX14 4RN

and by Routledge
711 Third Avenue, New York, NY 10017

Routledge is an imprint of the Taylor & Francis Group, an informa business

© 2019 David Clapham

The right of David Clapham to be identified as authors of this work has been asserted by him in accordance with sections 77 and 78 of the Copyright, Designs and Patents Act 1988.

All rights reserved. No part of this book may be reprinted or reproduced or utilised in any form or by any electronic, mechanical, or other means, now known or hereafter invented, including photocopying and recording, or in any information storage or retrieval system, without permission in writing from the publishers.

Trademark notice: Product or corporate names may be trademarks or registered trademarks, and are used only for identification and explanation without intent to infringe.

British Library Cataloguing-in-Publication Data
A catalogue record for this book is available from the British Library

Library of Congress Cataloging-in-Publication Data
Names: Clapham, David, 1951– author.
Title: Remaking housing policy : an international study / David Clapham.
Description: Abingdon, Oxon ; New York, NY : Routledge, 2018. | Includes bibliographical references.
Identifiers: LCCN 2018004697 | ISBN 9781138193932 (hardback : alk. paper) | ISBN 9781138193956 (pbk. : alk. paper) | ISBN 9781315639086 (ebook)
Subjects: LCSH: Housing policy.
Classification: LCC HD7287.3 .C575 2018 | DDC 363.5/561—dc23
LC record available at https://lccn.loc.gov/2018004697

ISBN: 978-1-138-19393-2 (hbk)
ISBN: 978-1-138-19395-6 (pbk)
ISBN: 978-1-315-63908-6 (ebk)

Typeset in Galliard
by Apex CoVantage, LLC

The book is dedicated to my grandson Tipene Narayanan who brings joy to everyone.

Contents

Acknowledgements	viii
1 Introduction: why housing policy needs to be remade	1
2 Housing policy	11
3 Housing regimes	24
4 Making the market	41
5 House and home	65
6 Neighbourhood	83
7 Housing supply	101
8 The distribution and affordability of housing	118
9 State-provided housing	142
10 Homelessness	159
11 Environmental sustainability	178
12 Neoliberalism and beyond in housing policy	192
Bibliography	209
Index	220

Acknowledgements

I would like to thank the many students who have explored with me the ideas put forward in this book. Three people have read earlier drafts of the manuscript and offered useful comments namely Bo Bengtsson, Rebecca Chiu and Keith Jacobs. They deserve many thanks for taking the time to do this despite their busy schedules. The staff at Routledge, especially Catherine Holdsworth, have been really helpful throughout the long process, and I would like to thank them.

I am grateful to Penguin books for permission to use material in Case Study 6.1 and to Routledge, AHURI and BSHF for permission to quote material from their publications. Ivan Tosics provided the photograph in Chapter 4. Many of the case studies are based on existing publications which are named as sources in the text. This case study material has been changed and edited to fit the context of the present book. Readers should refer to the original source for more information.

No book of this ambition can be completed without substantial time and effort, and this impacts especially on close family. So, I am particularly grateful for the love and patience of my partner Ravi and the understanding of all the children, Lizzy, Hannah, Belinda, Kieran and their partners.

Chapter 1

Introduction
Why housing policy needs to be remade

The housing problem is a recurring phenomenon in many countries. The inability of some people to achieve the housing situation that they expect or wish to achieve, whether because of a shortage of appropriate supply of the right houses in the right places, or the inability to be able to afford to pay for those houses, is an important political issue throughout the globe. The housing problem is shown in sharp relief by the number of people across the world who do not have adequate shelter or anywhere to call home. Although in most countries housing is primarily produced and distributed through market mechanisms, there is often pressure on governments and local authorities to devise policies and programmes to deal with housing problems. However, the primacy of the market can make it difficult to devise effective mechanisms to achieve public policy goals. Therefore, the main aims of this book are to examine some of the major objectives of public policy in the field of housing and to evaluate evidence on how effective these mechanisms are in different circumstances. This is intended to provide an analytical framework for policy-makers and commentators in order for them to be able to analyse policy mechanisms in a particular country, as well as providing a context of the experience and impacts in other countries that can serve as examples of possible future scenarios. It is hoped that this analysis will result in a re-instatement of government intervention in housing as a legitimate activity that can improve housing outcomes for many people.

The book is intended for an international readership, and so discussion is not centred on one particular country which is common in most housing policy books. Six countries are used in the book to provide examples of housing problems and attempts to deal with them and the countries are UK, USA, Australia, China, Sweden and Argentina. The justification for the choice of these countries is given in Chapter 3. It must be stressed that the book is not a comparative analysis of housing in these countries, but uses their experience to illustrate general points. They are used to derive case studies and examples that shed light on the general issues discussed and may provide food for thought for readers.

Part of the motivation for writing the book is concern at the general trend in many countries for governments to withdraw from responsibility for outcomes in the housing field and to accept market processes and outcomes under the influence of neoliberal ideology. A recurring theme of this book is that the neoliberal ideology when applied to housing is fundamentally flawed in a number of respects. First, it is based on a misconception of the nature of housing as a commodity, as is shown in Chapter 4. Second, the neoliberal ideology, when implemented, leads to outcomes that are unsatisfactory for large elements of the population. This is partly because neoliberal misconceptions about the nature of housing and housing markets mean that the desired and predicted outcomes often do not and indeed cannot materialise. In other words, the ideology cannot achieve the objectives held out for it and desired by those who implement it. But it is also because the implementation of the ideology leads to hardship for substantial portions of the population that governments either seem not to be concerned about or find difficult to prevent whilst adhering to the ideology.

It is not the intention here to provide a holistic and coherent alternative to neoliberalism, as a number of steps need to be undertaken before embarking on such an endeavour. For example, the increasing dominance of neoliberalism means that there are not widespread examples of alternative practices that can be evaluated and prescribed. In any case, it will become evident as the book progresses that it is vital in charting a way forward for housing policy to step outside the neoliberal ideology and to enact structural and holistic reform. Therefore, the contribution of this book is to demonstrate the outcomes of the neoliberal housing regime, to give examples of alternative practices where they exist, and finally in the conclusion to sketch out the beginnings of an alternative regime that takes as its base the concept of housing as a social right and places emphasis on the two related objectives of equality and sustainability.

There is some dispute about the meaning and value of the neoliberal label (see, for example, Birch, 2017), and it is true that 'neoliberal' is rarely a label used by people and politicians themselves. Without doubt it can mean many things, and it has become a catch-all phrase to describe many phenomena, but progress can be made by breaking down the concept. Ideologies can be seen, following Freeden (1998), to comprise flexible and shifting structures of essentially contested political concepts, whose function is to simplify and control the proper meaning and structure of political thinking – in other words, to constitute the discourse that shapes understanding. Neoliberalism is seen here as a discourse, a set of powerful ideas that are sustained and used by powerful agents to further their interests. Birch (2017, p. 68) argues that

> economic ideas help to frame political-economic policy problems, help to identify possible solutions to those problems, and help to establish the effects of those solutions. As such, ideas come to represent a series of goals or objectives to which social actors orient themselves and their understandings of the world.

Freeden's approach identifies three levels of concepts within an ideology: the core, the adjacent and the peripheral. Concepts situated at the core of an ideology are the most general and the most fundamentally important, while the adjacent concepts flesh out the general meanings with greater richness and nuance. The main focus here is on the peripheral concepts that are of less fundamental importance to the structural integrity of the ideology and can include 'perimeter' concepts which are theorised as being 'specific ideas or policy-proposals rather than fully fledged concepts' (Freeden, 1998, p. 78). Peripheral concepts are the means by which a political ideology interfaces with the political world and which enable the implementation of the general ideas into specific policy realms such as housing. However, it is difficult in practice to separate out the different levels and what is presented below is a mixture of the different levels, although their specific application to housing is drawn out.

There is some disagreement about the core concepts of neoliberalism. For Wacquant (2009, p. 307), neoliberalism entails 'the articulation of four institutional logics': promotion, typically via economic deregulation and invariably in the name of efficiency, of markets and market-like mechanisms; welfare state retrenchment; propagation of a trope of individual responsibility and the entrepreneurial self; and an expansive and intrusive penal apparatus. However, there is some dispute about a number of these. For example, Birch (2017) and Davies (2017) show that there have been different understandings of the nature and importance of markets. For neoliberals of the Chicago School of Milton Friedman and others, markets should be free of government 'interference' and pure, that is competitive. This view held sway up to the 1960s but was superseded by a view that corporate monopoly was not as bad as thought in delivering effective economic outputs and that anyway it was 'a transitory phenomenon that will ultimately be eroded by market forces' (Birch, 2017, p. 110).

The centrality of markets to the neoliberal discourse has been recently queried from a direction that relates very closely to housing. It is argued that from the 1970s and 1980s there has been a move away from a concern with markets and competition towards financial capitalism or rentiership, which

> involves the appropriation through government fiat (e.g. laws, regulations, standards) monopoly rights (e.g. location), and organizational reconfigurations (e.g. mortgage securitization). It is not, in this sense, about the production or creative generation of value or wealth; rather, it reflects a reliance on unearned income and speculative wealth.
> (Birch, 2017, p. 143)

Following Sayer (2015, p. 50) 'A person who derives unearned income from ownership of existing assets or resources is known in political economy as a rentier'. The proportion of the national income derived from rentiership has increased substantially in many countries and housing is at the heart of this. The primacy of markets has been superseded by a belief in the rights of the investor to choose the best investment and so to allocate resources efficiently.

Also, there is controversy over the centrality to neoliberalism of welfare retrenchment or 'austerity' as it has been more recently known (see Blyth, 2013). In many countries perceived to have neoliberal governments (for example, the UK in the 1980s under Thatcherism), total public expenditure increased. However, there was a shift in the objectives of this expenditure as well as large distributional changes in benefits between different individuals and groups. It can be argued that the Global Financial Crisis (GFC) in 2008 has led to a change here with much more emphasis on public expenditure reductions in many countries in order to right the public finances after the demands of saving the banking system. Individual responsibility and the concomitant freedom and choice are less controversial elements of neoliberalism with the focus on individualism and liberty, although there may be differences in how this is defined. The extension of the penal apparatus is usually conceived as being the means of perpetuating these elements of the neoliberal discourse.

Therefore, the central concepts of neoliberalism are contested, loosely defined and sometimes contradictory. But the ideas remain powerful in political discourse and strongly influence policy in many countries. They are reinforced by the tenets of neo-classical economics that shares many similar assumptions about the centrality of markets and of individualism. As we shall see in the following chapters, much housing policy is underpinned by evidence derived from neo-classical analysis based on unrealistic assumptions of the nature of housing and the housing market.

The key elements of neoliberalism are too general to offer more than a very vague guide to policy. Therefore, the focus here is on the concepts that relate more strongly to the policy domain, and five are identified that have been important in housing policy. The first is *privatisation*, which is defined as moving housing assets and ownership from a state to a private sphere, such as through the sale of council houses in many countries.

The second concept is *marketisation*, which is defined as the increasing scope of market exchange, market relations through competition and market behaviour and thinking by agents in the housing regime. Davies (2017) argues that, while neoliberal states have extended and liberated markets in certain areas (for instance, via privatisation and anti-union legislation), the neoliberal era has been marked just as much by the reform of non-market institutions, so as to render them market-like or business-like. In general, this is associated with the implementation of 'New Public Management' to make the public sector work to the same rules and processes as the private sector by installing competition and market thinking. A good example of the approach is the requirement now made of Swedish Municipal Housing Companies that they operate according to market principles (although there is some dispute about what this means in practice).

The third concept, which is closely related to the other two is *commodification*. The emphasis on markets and competition in the neoliberal paradigm has led to housing being commodified, that is seen as predominantly a traded commodity which is valued for its financial status, rather than as a human right or a product valued for its use rather than its exchange value. The distinction between housing as a marketed commodity and as a basic human right is an important one that will resonate throughout the book as we see the consequences of the commodification and neoliberalism more generally.

The fourth concept is *financialisation*, which is defined as the growing influence within housing of its financing structures and agencies and an increasing integration of housing into the global financial markets (see Chapter 4) which relates to the debates about rentiership outlined earlier.

The fifth concept is *individualisation*. An example is the sale of council houses in the UK (and in many other countries), which took a communal good, the surplus built up in terms of the equity in the sector which was being used to subsidise new development, and individualised it by giving it to individual households through the discount given to right-to-buy purchasers.

In this book, the neoliberal ideology is understood as these five concepts that have been applied to the housing sphere. As we shall see in Chapter 3, these have been constituted into a neoliberal housing regime that has shaped the nature of housing processes and outcomes in many countries that are investigated in the following chapters. A housing regime is defined as 'the set of discourses and social, economic and political practices that influence the provision, allocation, consumption and housing outcomes in a given country' (see Chapter 3). The overall impact of the neoliberal housing regime is assessed in the conclusion.

The result of the application of these neoliberal concepts and the adoption of a neoliberal housing regime has been an increasing polarisation of housing circumstances across different social classes and ethnic groups, reflecting the growing wealth and income inequalities that many countries have experienced under financial and economic globalisation. At the same time, the uneven processes of liberalisation have left many governments holding policy tools and mechanisms that may not be perceived as effective in meeting their goals. One example may be social housing that was originally devised to offer a comfortable and well-maintained home for a wide spectrum of people, but which has become a small residual sector reserved for the poor and ethnic minorities that serves to segregate them economically, spatially and socially from the rest of the population. Therefore, its use as a mechanism for achieving social inclusion and offsetting inequality has been substantially compromised in some locations. Another may be regulation of the private rental sector that was devised to improve security and affordability for low-income households, but which is increasingly perceived as restricting market processes and restraining housing supply. The perceived 'failure' of these mechanisms leads to more pressure to reduce government involvement and to further 'unleash' market processes. But as we shall see in the following chapters, there may be a place for these mechanisms in a refocused housing regime.

The uneven withdrawal of the state has led in some countries to the creation of what Christophers (2013) has labelled a 'monstrous hybrid' with a mix of market and state mechanisms that produces an inefficient and unstable outcome. His analysis was of the Swedish housing regime, but much the same can be said of Britain and other countries that have taken the neoliberal path. Although all regimes can be seen as hybrid in the sense that there is a mix of state and market influence, Christophers argues that they have become monstrous because the two elements are contradictory and result in outcomes that can be seen as inefficient and harmful to sectors of the population. Therefore, the time seems ripe to take a more fundamental look at the aims of housing policy in these countries and to examine whether existing policy mechanisms are useful for purpose, or need to be refashioned or replaced.

There is not a clear and simple link between the central and peripheral ideas of neoliberalism identified here and housing (or any other) policy. The ideas are mediated by existing social institutions such as the state, the family and corporations. Therefore, there is what has been termed *path dependency* in which the existing institutional structure sets the terms for the extent and form of change. This concept is introduced more fully in Chapter 3, where we examine the reasons why policy will vary between different countries.

Some of the most important frameworks for the comparative study of housing such as 'path dependence' and the 'varieties of residential capitalism' (see Chapter 3) argue that housing policy is primarily shaped by the institutional, economic and financial structures that surround it. Institutional structure here is used to indicate not just the organisations that build, distribute and manage housing, but also the social practices or patterns of social interaction between people or agents (including housing consumers) that represent the housing regime. Adherents of the path dependence approach argue that the character of this institutional structure patterns the nature of the housing regime in different countries and that change in this regime is slow and gradual because of the inertia and vested interests of the institutions within it. However, it is accepted that there are certain critical junctures at which governments choose particular paths and where radical change may take place, although always constrained by the existing institutional structure. Advocates of the 'varieties of residential capitalism' approach argue that change is brought about through changes in the structure of capitalism and also use the concept of critical junctures to highlight the times that this change takes place. Are we at a critical juncture in many countries? Certainly, there is a growing disquiet in many places about the direction of current policies and the 'monstrous hybrid' that neoliberalism has left behind, and that has contributed to the seemingly intractable growth of housing problems such as homelessness. The Global Financial Crisis of 2008 has highlighted pressing housing problems in many countries and the seeming inability or unwillingness of national governments to deal with them.

Nevertheless, the rise of neoliberalism in housing has resulted in housing policy itself becoming unpopular in some quarters, with the legitimacy of the market being given precedence over state intervention, which is seen as involving inappropriate interventions that distort the market. Hayek (1944) saw markets as the triumph of economics over politics with the hidden hand of market forces taking over from the collective deliberation and decision-making of politics. Such a view sees no role for the collective and conscious direction of a housing policy because this is seen as inevitably leading to the demise of freedom and the imposition of serfdom. Following this paradigm, when housing problems are evident, the cry from neoliberals and their followers is for even less government 'interference' and the 'freeing up' of the market, although there is little evidence that this will lead to more desirable outcomes. In addition, this view hides the important role of government in setting the legal and institutional structure of the market (see Chapter 4). As we shall see in later chapters, not all housing markets are the same in different countries as their form is dependent on governments. Therefore, all countries can be said to have a housing policy even if it is restricted to setting the rules of the game for the market.

It is not assumed in what follows that all governments in different countries will agree on what desirable housing outcomes are. In recognition of this, an important approach in housing studies follows the work of Esping-Andersen (1990) in identifying different welfare regime types reflecting different political ideologies and policies towards housing and welfare services. Esping-Andersen identified three different regimes (social-democratic, conservative and liberal), but subsequent analyses have added to this list (see Chapter 3). Building on this work the concept of a housing regime is identified and defined in Chapter 3, and the important elements by which regimes may differ are shown. In addition, the housing regimes of the six countries used as examples in this book are described. Any conception of a universal, ideal housing policy is based on a set of political ideologies grounded in moral and ethical principles. The aim in this book is to lay out these principles for inspection and to examine their resultant impacts when implemented into housing regimes.

Despite the focus on the activities of government, the assumption in this book is that housing is primarily provided through market mechanisms as it has been in almost all countries. The only exceptions are probably some eastern European countries between the end of the Second World War and the fall of the Berlin Wall and the overthrow of communism. In those years, an Eastern European housing regime operated in some countries that relegated market mechanisms to unofficial and minor roles, with the state being the primary producer, owner and manager of housing, which was envisaged as part of a social wage and provided for only a small cost to the majority of the population (see Clapham, 1995). However, this regime has now been superseded in almost all countries and replaced with market dominant forms. Therefore, the market is the predominant form for the delivery of housing in almost all countries, and the role of government is to set the conditions for market functioning and to intervene to achieve desired outcomes. The role of the market in housing is rarely questioned in the way it is for other services and commodities. For example, Sandel (2013) has shown how market relations have taken a greater role in many aspects of life and have resulted in a set of moral relationships based on economic incentives. Foucault (2008) points to the way that markets need consumers as well as producers and that the state is active in making people culturally to become individualised consumers who give primacy to exchange value over other forms of value. In the following chapters, it will become clear that many of the housing problems that face governments have, at their heart, the difficulties that the adoption of this commodification approach causes in terms of attitudes towards housing, as well as the outcomes generated. The dominance of market relations fosters a perception of housing as being a means to generate or store wealth rather than as a means of shelter and as a place for pursuing family life. An alternative viewpoint sees housing is a universal human right rather than a traded commodity and which, therefore, should be allocated on the basis of need rather than ability to pay. Nevertheless, if governments have held this alternative set of moral values they have decided to achieve them by regulating or shrinking the role of the market rather than removing it. Housing is truly the 'wobbly pillar' of the welfare state (Torgerson, 1987) in which the values of social welfare have never been fully dominant.

The primary focus of the book is on the relationship between the state and the market. However, it is recognised that there are also two other sectors of activity that do not fall into these two categories and may be important in housing in some countries. One of these categories is family, and a 'family' regime that has been identified as important in some countries such as in Southern Europe. Here young people will often build their own house on family-owned land rather than have recourse to a rental sector or developer provided private housing market. A second category is civil society, involving community or collective action. For example, housing co-operatives may produce and manage housing collectively outside both the state system and the private market. Individual activity such as squatting or self-building may also come into this third-sector category. Of course, this sector also may come under state regulation and may have to use the legal structure established through state legislation. But it is also important to note that activities such as squatting or protests about housing regeneration may form 'social movements' (see Madden and Marcuse, 2016) that may influence the policy-making process and change public perceptions of issues.

In order to achieve the aims of this book, it is necessary to examine housing policies in relative isolation from their social and economic context. For example, the approach used here is to examine the outcomes of policy mechanisms (such as the provision of public housing or the imposition of rent controls on private renting) across different countries and different contexts. Some authors in the path dependence or varieties of capitalism approaches would argue that housing policy is crucially dependent on the context within which it is situated and that there are certain economic and social forces (such as globalisation see Clapham, 2006) that are the determinant factor in housing policy. Undoubtedly this context is important and must be borne in mind in all that follows. However, it is important not to under-emphasise the importance of the political choices made by governments that exist in any context and which are foregrounded by the welfare regimes approach. The emphasis in the book will be on the choices available to governments, although it is recognised that these are crucially influenced by the context within which governments are operating. Therefore, care will be taken to ensure that contextual factors are taken into account in the evaluation of housing mechanisms in the following chapters. However, the main aim of the book is to provide a framework for policy-makers and commentators to be able to analyse policy mechanisms in a particular country and to provide a context of the experience and impacts in other countries that can serve as examples of possible scenarios. It is certainly not assumed that there is one policy answer for all countries to adopt, although the obvious drawbacks of the neoliberal approach that will become evident in the following chapters, mean that this approach is not one to be advocated here. Indeed, it is argued that the dominance of this perspective needs to be challenged if answers are to be found to pressing housing problems.

Outline of the book

Chapter 2 starts by defining the term 'housing policy' as used in this book and examines the different forms or mechanisms that are available to governments in order to achieve policy aims. It continues by reviewing different approaches to understanding the way that policy is made. The approach adopted in this book is the discursive approach in which the focus is on the language games, coalition-building strategies and power activities that shape policy outcomes. Evidence can help shape these games, and so the information provided in the book is intended as a contribution to the debates and negotiations around housing policy, rather than definitive answers to particular issues. Given this approach, the chapter focuses on how to judge a good housing policy. It is argued that policy should be evaluated in terms of the outcomes experienced by 'dwellers', that is people who live in houses or who could potentially live in a house. It is argued that housing outcomes are subjective and may vary between individuals, and so the concept of subjective well-being is introduced as the basis for judgements, and this is usually gauged through measures of life satisfaction. Therefore, it is argued that the test of a good housing policy is whether it increases the life satisfaction of dwellers. However, it is recognised that there may be trade-offs between different groups. One policy may help one group at the cost to another. Therefore, well-being needs to be considered in the context of moral judgements of social justice that vary between different political ideologies.

Chapter 3 introduces the concept of a housing regime that is used to show that there are differences in the way that countries approach the objectives of housing policy and that this factor has to be borne in mind in the discussions that follow. The chapter describes the nature of a housing regime and sketches out the constituents of the housing regimes referred to in the book such as the neoliberal regime. The chapter then introduces the six countries used for case studies in the following chapters. It is stressed that the book is not a comparative analysis of these countries, but uses examples in order to illustrate particular points or themes and to highlight differences and similarities between countries. It is not assumed that what works in one country will work in another with a different context.

The book continues by using the framework derived in the previous two chapters to examine specific elements of housing policy based on the objectives that are sought. Chapter 4 describes the way that governments set the basic foundations of the housing regime by 'making the market'. The chapter starts with a review of the particular features of housing as a commodity or economic good that make it difficult to deliver through market mechanisms, such as its complexity, the incidence of significant externalities, the importance of information asymmetries between consumers and producers and exchange professionals such as estate agents, and the divergence between individual and social benefits and costs. These factors form the conditions and the justifications for government interventions that are considered in later chapters. The chapter continues by focusing on the market itself and argues that markets have to be constructed and maintained by governments. Therefore, all governments undertake the structuring of a functioning housing market, although the way they undertake this differs considerably and, therefore, so do the types of markets that result. The chapter then focuses on major forms of market regulation such as the construction of housing tenures and exchange mechanisms as well as the regulation of financial institutions and processes. It is argued that the financialisation that is a feature of many housing regimes is leading to difficulties for governments in employing appropriate policy tools to alleviate the problems. This is followed by a discussion of the problems of increasing house prices and volatility that occur in many countries and especially those with neoliberal housing regimes.

Chapter 5 uses the concept of affordance to examine the relationship between residents and their houses and the ways that governments may intervene in this. A key theme is the way that housing has both meaning and utility elements that are both important in creating home. Apart from a few defined situations involving clear health triggers, the well-being of dwellers is personal and related to the meaning and status aspects of a house. There are a number of justifications for government involvement in the quality of housing that relate to the harm caused to dwellers or various forms of externalities such as harm to others or the importance of wider social objectives such as sustainability and the needs of future generations. However, the subjective nature of the well-being derived from houses makes it difficult for governments to design appropriate tools and methods of intervention or to assess the impact of common mechanisms such as the application of building regulations or public health standards. The chapter concludes by drawing attention to the positional nature of housing and the impact this has on dweller expectations of standards. It is vital in making policy to understand how these expectations are formed and the impact that the inequality of housing outcomes has on them.

A key element of the quality of houses is their location and their neighbourhood environment. Therefore, Chapter 6 examines neighbourhoods and the issues of social segregation and social cohesion that can be the focus of government housing policy. The key questions in this chapter are whether it matters where people live and how governments can alter this. Discussion focuses around ideas of balanced or sustainable neighbourhoods, and the main mechanisms that governments use to achieve these aims such as slum clearance, neighbourhood renewal and the use of urban planning powers will be considered and evaluated. It is concluded that this is a problematic area for housing policy that is beset by dilemmas as to the appropriate scale of intervention and decisions about who benefits and loses from the intervention. The chapter points to the neoliberal focus on public sector neighbourhoods as part of the privatisation agenda and the impact this has on local residents.

Governments may want to intervene in the market to influence the amount of housing built, and so Chapter 7 considers housing supply systems and why housing shortages may occur. It describes how governments can decide whether there is an undersupply of housing and the mechanisms they can use to intervene to increase supply. Examples may include the urban planning system, supply-side subsidies, taxation and direct provision. It is argued that neoliberal housing regimes are characterised by increasing concentration of large firms in housing development and that financialisation has meant that they are primarily focused on short-term share value, often to the detriment of housing supply.

Chapter 8 focuses on the patterns of the distribution of housing that arise from different housing markets and their relationship to the distribution of income and wealth. The concept of 'affordability' is defined and possible bases for government intervention in affordability issues such as the concepts of fairness and equality are discussed. The mechanisms that governments can use to alter the distributional pattern are described and evaluated, including demand-side subsidies, taxation policies or other financial mechanisms such as the granting of secured finance. It is argued that the adoption of a neoliberal housing regime has led to increased housing inequality and that policy mechanisms employed to deal with affordability problems often result in increasing long-term difficulties.

Chapter 9 focuses on one particular, but important policy intervention, which is the provision of housing by the state. Direct provision, usually of rented housing, can meet many of the objectives that the state may hold in housing, but it is one method of intervention that varies considerably between countries and has been in decline in the neoliberal era. The chapter discusses the different forms that this housing may take and the evidence of its success and failure in different countries and at different times. Examples are used to show the context that is needed for this form of provision to meet the objectives held for it.

The focus of Chapter 10 is homelessness, which is the most extreme form of a lack of housing. Governments may adopt a number of philosophies and mechanisms to prevent and deal with homelessness and the chapter describes these and evaluates their effectiveness. Two common models are what are termed the 'staircase' and the 'Housing First' approaches. The chapter reviews the evidence on the outcomes of these forms of intervention. It is argued that high rates of homelessness are endemic in the neoliberal housing regime and that the characteristics of the regime make dealing with homelessness problems very difficult to achieve. Therefore, it is argued that strategies to reduce and cope with homelessness need to focus on general housing regime factors, particularly those that impact most strongly on the housing situation of low-income households.

Chapter 11 examines the issues of sustainability and climate change and their possible impact on housing. It is argued that this is one of the most important issues facing housing policy, but also one of the most intractable and complex, with implications throughout the housing regime. The chapter reviews the impact of climate change on housing and *vice versa*, focusing on issues such as energy use and resilience to climate changes. The chapter draws attention to the importance of social practices in housing driven by consumerist values in influencing the impact of sustainability policy. It is argued that the need for environmental sustainability poses an existential threat to neoliberal housing regimes that are based on ever-increasing consumption driven by status concerns exacerbated by high levels of inequality and can only offer minimal housing standards to low-income households through extensive resource use. However, sustainability challenges both the decision-making processes of the market as well as utilitarian concepts such as subjective well-being that have been used in this book to evaluate the impact of housing policies.

Perhaps more than any other area considered in this book, the existence of sustainability problems makes the case for a strong housing policy that takes into account the needs of future generations.

The aim of Chapter 12 is to draw conclusions about the form and impact of neoliberalism on the shape of housing policy in different contexts based on the analysis in the preceding chapters. It is argued that many of the housing problems that confront governments and impact on the lives of many, particularly vulnerable households, are the direct outcome of the neoliberal ideology and are endemic to it. In conclusion, the chapter looks for a way forward in housing policy that seeks to break free from the neoliberal ideology.

Chapter 2

Housing policy

The primary aim of this book is to provide evidence to enable change in housing policies. In order to achieve this, we need to define what is meant by housing policy and what forms it takes. What are the key policy instruments that governments can use to achieve their objectives in housing? We then need to examine how housing policy is made in order to establish how evidence on policy outcomes, such as that provided in this book, can influence the process. There are a number of ways of viewing the policy-making process, but we adopt a discursive approach that sees information as contributing to language games that are played by policy agents with differential power positions. The focus is on the language games, coalition-building strategies and power activities that shape policy outcomes. Evidence can help shape these games, and so the information provided in the book is intended as a contribution to the debates and negotiations around housing policy. The chapter then turns to consider the kind of evidence that is useful in the policy process. The key element here is to understand how housing policies should be evaluated and what constitutes a successful policy. It is argued that housing policies should be judged on the basis of the outcome for dwellers and that the concept of subjective well-being is useful in assessing this. Finally, the chapter notes the need to see housing outcomes in the light of political values and ideologies that are the focus of Chapter 3.

What is housing policy?

First, it is important to define what we mean by housing policy. In this book, housing policy is taken to mean 'any action taken by any government or government agency to influence the processes or outcomes of housing'. A number of items in this definition need to be explained in more detail.

There are many influences on housing policy, and the term 'governance' used in contrast to the term 'government' shows the acceptance that policy is made through networks consisting of a number of public, private, voluntary and hybrid organisations. Therefore, it can be argued that policy is made by all of these bodies, and so research aimed at policy relevance can be directed at any of them. This is accepted, but at the same time, the assumption is made here that the state holds a special place in this network, and in many countries, it is expected that the state takes ultimate responsibility for the outcomes of the housing system. Nevertheless, the argument of the chapter holds even if it is held that government agencies do not hold this special position and policy relevance can be assessed in relation to a wide range of different organisations. Therefore, policy relevant research (which this book aims to produce) is defined as research that impacts on the policy process in some way and is used by any agency involved in that process.

Government or government agency is taken here to mean any agency accountable to elected politicians or agencies owned and controlled by the government. Of course, this is a very difficult aspect to define, and there is substantial ambiguity. For example, there are different views about whether British housing associations are public bodies or not. They receive subsidy from and are regulated by government, but legally, they are charities or companies responsible to their trustees, owners or shareholders. They are clearly semi-public bodies that have both private and public elements.

Policy interventions may differ in their form and their scale. In terms of geographical scale, housing is global, national and local in its reach. Little exists in terms of global policy, although the United Nations has intervened by commenting on national housing policies and announcing that housing is a basic human right. In addition, some global institutions such as the International Monetary Fund (IMF) or the World Bank may actively intervene in the housing policies of some countries. For example, Murray and Clapham (2015) show the impact of World Bank policies on housing regimes in South American countries. In Europe, housing is not a competence of the EU as it is a devolved function to national governments, but housing ministers of the member countries meet regularly and exchange ideas and information and some Non-Governmental Organisations (NGOs) such as Feantsa organise at the European level in their attempts to change housing policies. In addition, many EU policies on taxation, welfare or the economy have important impacts on housing even if they are not directly housing policies aimed primarily at housing outcomes.

Housing policy is usually considered to be essentially a national concern. However, in countries with a federal structure such as the USA or Australia, the states will have a major role in policy alongside the national government, and in the UK, housing is one of the devolved functions to the constituent countries of Wales, Scotland and Northern Ireland. However, policies that have a major impact on housing such as income support or economic policies are not devolved, and so national parliaments or assemblies can find themselves without major policy tools and can have their efforts undermined by national government policies in retained fields.

In most countries, local governments are involved to a greater or lesser extent in the making and implementation of housing policy. The usual pattern is for central government to set the national funding and legislative framework whilst allowing local authorities to decide on the specific policy mechanisms to use in their area depending on the particular situation they face, although countries differ in the extent of freedom and funding that local government may have.

Therefore, government agencies can exist at many different levels. Some have responsibilities at the national level whereas others may be local municipal agencies that have local accountabilities and duties, although this may be within a national framework or financing structure. Other agencies may be national in scope, but be accountable at arms-length to central government. An example may be a national housing bank that operates under a statutory framework, with its strategy and finance emanating from central government.

Whatever the scale of intervention, governments and government organisations have a number of types of mechanism that they can use.

The first is *regulation*. This could involve setting the limits of action of private actors and institutionalising social practices in the housing market. A common example would be regulation of the private, rented sector that could involve controlling rents, constraining the activities of lettings agents and providing security of tenure to tenants.

The second form of intervention is *direct provision*. This may mean the national or local government or state agency directly building housing for sale or, more usually, for rent. Government may not undertake all of the stages of housing development in that it may contract out actual building work to private contractors or work with private housing developers. The example of direct provision most often encountered is the provision of rented housing, but the state may provide other housing services such as accommodation and support for homeless people.

The third form of intervention is through the provision of *finance or subsidy and taxation*. These are three different activities but are grouped together here because they are often compared and the analytical process to evaluate them is similar. All three activities involve changing the incidence and size of costs and benefits to the different parties involved. Government may make available grants or loans to individuals or organisations to achieve particular objectives. Payments to individuals may be to allow them to afford housing they otherwise would not be able to or to alter their perceptions and attitudes and behaviour towards, for example, reducing energy use in a home or repairing the home. Grants to organisations such as private builders and developers may be designed to increase the supply of housing or to change its nature by, for example, improving standards of energy efficiency. Provision of finance may constitute a subsidy in that people or organisations are given resources not available to others and which can be argued to subvert or distort market processes. Taxes may be levied on particular elements of the housing process or tax reliefs made available in certain situations.

The fourth form of intervention is through the provision of *information or guidance*. In Chapter 4, attention is drawn to the information asymmetries between participants in the buying and selling of houses. Therefore, government may act to provide some of this information or encourage or compel another party to provide it. Information and guidance may also be provided to vulnerable individuals who have problems that mean that they require help from social work or other professionals to be able to participate successfully in the housing process.

The fifth form of intervention is setting the patterns of *accountability* for organisations in housing. In other words, government can define the relationships between the parties involved in housing. An example may be setting targets or monitoring procedures to ensure that social housing agencies have to consult their tenants in particular circumstances.

Sixth, governments are often active in the field of discourse around housing. In other words, they are important in setting the terms of discussion and debate and *defining issues and problems*. For example, countries differ in their attitude towards homelessness with some seeing it as a result of imperfections in the housing market and others as a personal failing by the homeless people themselves. Whichever discourse is dominant and accepted by government frames debate on the issue and defines the way the problem is defined and the actions taken to deal with it. Some policies will have a strong symbolic element in relation to political beliefs and ideologies. For example, owner-occupation is sometimes thought to symbolise the importance of private property rights and to lend legitimacy to market relations and minimal state intervention.

The seventh element is *non-intervention*. Following the debate about the importance of non-decision-making as a form of power emphasised by Lukes (1974), it is argued by Doling (1997) that choosing not to intervene in the market in general or in specific circumstances is a housing policy in its own right that will have specifiable outcomes.

The definition of housing policy used in this book focuses on interventions designed to impact on the housing system. However, cognisance will have to be made of two other forms of intervention. One form is interventions in housing that are aimed at objectives in other fields. Examples may be housing finance policies that are aimed at influencing the national economy or house building regulations that are set to achieve environmental goals. The second form is interventions in other sectors that influence housing. An example of this may be social security policy that will influence the capacity of individual families to pay for housing. Both of these policy categories may be very influential in housing outcomes for many people and so will also be considered in this book, although space considerations limit the coverage.

Making housing policy

In order to understand the role of evidence, as provided in this book, in the formulation of housing policy there is a need to examine how housing policy is made. There are many different approaches to guide governments and other agencies to make housing policy. The traditional schism is between approaches that focus on a rational analytical process and ones that assume that policy-making is an inherently conflictual process that is essentially political in nature. Of the latter, three variants are discussed here, the incremental political approach, structural explanations and the discursive approach.

The rational approach

Analytical approaches tend to see policy-making as a process involving finding information and making decisions based on rational analysis. An example of the approach would be a 'rational decision-making' process that follows a number of steps (see Hogwood and Gunn, 1984). The first is scanning the environment and gathering information about circumstances and problems; the second is setting objectives in relation to the perceived problems; third is identifying different options for achieving these objectives; fourth is appraising the options and deciding which one achieves the objectives in a most cost-effective way; the fifth stage is implementation; and finally the outcome of the policy is evaluated to assess what impact it has had on achieving the objectives set for it. This process is usually seen as on ongoing one in which the evaluation of the outcome feeds in to the first process of environmental appraisal and is used to re-assess and reset objectives and means of achieving them in a continuing cycle. Many forms of this process are advocated, but share the basic tenets of rational analysis in an ongoing cycle. One criticism of this approach is that it is impossible to achieve in practice as the constraints of time, resources and knowledge mean that it is very difficult – if not impossible – to foresee the costs and benefits of different means of achieving a particular policy, and so analysis is usually going to be limited in some way. Simon (1947) adopted the concept of 'satisficing' to describe how policy-makers try to be as rational as possible given the constraints they face. Etzioni (1986) put forward a mixed-scanning approach in which he argued that the best approach was for policy-makers to continue in an incremental manner for most of the time, but then at certain times (maybe at critical junctures) to take a more comprehensive and rational view to enable more radical change to take place if needed. Although both authors voice criticisms of the rational approach, they remain firmly within the rational category as they share many of its fundamental tenets such as the belief in the importance of 'rational' and value-free analysis.

For a rational process, there is an implicit positivist and empirical approach towards evidence. The search is for social facts that can prove the existence of certain phenomena or allow the judgement of the success of a particular policy. The rational approach would look to answer questions about the outcomes of a policy by saying that a good policy achieves the objectives laid down for it. But objectives may not be explicitly stated and not everyone may agree what they are either in general or for specific policies. Also, it is important not to focus exclusively on objectives as many policies have unintended outcomes as well as intended ones and these may be important in terms of their acceptability to politicians and people in general as well as the overall judgement of their worth. Also, many of the most important housing objectives relate to the condition, distribution and affordability of housing, and so measuring the success of policies depends on being able to identify and assess their impact on the housing outcomes of individual households. Therefore, evaluation has to focus on outcomes. In addition, the impact of housing policies may differ between individuals and groups.

The political approach

There are fundamental critiques of the 'rational' approach that place emphasis on the political elements of policy-making. Lindblom (1965) argues that rational policy-making is impossible in practice, and so policy-makers simplify the process by only making incremental changes in a kind of 'suck-it-and-see' manner in which change is monitored and, if successful, continued and, if not, discontinued. It is generally accepted that much of government policy-making in the real world takes this approach, but the debate focuses on whether this is desirable or whether rational policy-making should be more prevalent.

Lindblom's (1965) most fundamental critique of the rational approach is that policy is made through the political process that involves bargaining and negotiating between the parties involved. The parties may include government officials and departments, private sector interests, consumers, social movements, the media and any others with an interest in the issue concerned. He argued that a good policy was one that was agreed on by the parties involved in it. Agreement meant that the policy was more likely to be implemented and to be sustained. Criticism has focused on the assumption that all relevant parties that would be impacted by a particular policy are involved in the policy-making process and on the importance of imbalances of power between them in determining the final outcome. Some parties may be powerful, resourceful and well organised and so able to set the terms of the debate and enable adoption of their preferred solution whereas others may be powerless and disorganised. The awareness of this imbalance led to advocacy planning in which the concern was to help powerless people to enter the process and make an impact on policy.

As well as raising the question of power, Lindblom's focus on the political process raises issues around discourse and agenda setting. He argues that parties may have different views on the problems and the issues involved and that there may need to be agreement on what the problem is as well as how to deal with it once defined. The political approach to policy-making alerts us to the existence of different sets of objectives held by groups in society. Rather than accepting the dominant discourse and the objectives that flow from it, analysis should take into account the different perspectives and values that are implicit in alternative discourses.

So, if it is assumed that the policy process is a pluralistic bargaining between interest groups, those providing evidence of the success of housing policies can expect their findings to be debated and contested as part of ongoing negotiations. It could be argued that evidence can improve the outcomes by changing the nature of the bargaining involved by improving the information on which the debate is based. However, this approach still neglects the presence of different discourses and definitions of the problem to be addressed. If policy-making is about the reconciliation of different perspectives and discourses, then rational analysis may not help this process.

The political approach seems best suited to research based on a social constructionist perspective that emphasises the socially constructed and contested nature of reality. Research that elucidates and analyses different discourses held by different groups would help to inform policy by providing information for the bargaining process rather than providing a 'right' solution that would transcend any debate. Therefore, the focus in the research would be the elucidation of discourses and attitudes and perceptions. Analysis can highlight logical inconsistencies in any discourse and compare with other discourses and research evidence in terms of socially constructed 'facts'. The research may highlight agreements and differences between discourses and so help in the bargaining process.

The political nature of policy-making is recognised by many decision-makers today as evidenced by the large extent of research undertaken by pressure groups and 'think tanks' that share the same discourse as the decision-maker. Even where a rational analysis is undertaken, its findings may be subjected to political analysis through the lens of the decision-maker's discourse.

Structural approaches

There are many different structural analyses, but the focus here is on one of these that is relatively popular in housing research, and that is the Foucauldian perspective that focuses on issues of discourse and governmentality. Research in this tradition highlights the historical evolution of discourses and their impact in terms of the distribution of costs and benefits and the constraints on behaviour through the instruments of policy. Other structural approaches include critical realism and recently literature on the 'Varieties of Residential Capitalism' that seeks to show the impact of factors such as financialisation on the trajectory of different national housing systems (see Schwartz and Seabrooke, 2009; Aalbers, 2016). From these approaches, there have been major research emphases on the historical changes in policy over time and their underlying causes (see Lawson, 2006). But the approach also has a similar focus to the political approach as it emphasises the relationships between agency and structure and can draw attention to the contextual factors that influence behaviour. The emphasis on discourses and control mechanisms sheds light on the behavioural and symbolic elements of policy. Research of this kind may be of benefit to governments concerned to improve their policies and mechanisms, but it may also be important in changing public perceptions of dominant discourses by providing insight into the mechanisms used and their impact. The research may also be important to social movements and other resistance groups by providing alternative discourses and offering information on the impact of policies.

The discursive approach

The political and structural approaches offer important insights into the policy-making process and the place of evidence in this. The discursive approach is a useful way of bringing together these insights. Authors such as Healey (1997) and Fischer and Forester (1993) built on the political approach and put forward concepts of 'collaborative' or 'discursive' planning that recognise the importance of different discourses held by agents in the policy-making process and emphasise the importance of the process of communication and negotiation between them. Durnova et al. (2016, p. 36) consider policy-making as 'fundamentally a political activity pursued through beliefs, meaning and argumentation'. Discourse is at the heart of policy because it frames problems and the means through which actors interact. Policy solutions are not neutral instruments but have meaning. They argue that it is impossible to analyse public policies objectively and rationally but point to three stages in the analysis of policy. The first is the analysis of 'language games' that actors use to frame issues and, following Lindblom, to simplify decision-making. These language games can be transformed through critical response. In other words, the counter-arguments put by other actors or the results of empirical research have to be responded to, and this may result in change in the language and discourse. The second stage is the analysis of coalition-building strategies in which actors attempt to build support for their position. The third stage is the analysis of the power relationships in which actors seek to negotiate and impose their views in a conflictual arena. This stage coincides with the structural approach to policy analysis or what Durnova et al. (2016) termed 'critical policy analysis' For example, Foucault has used the term 'governmentality' to highlight the different strategies and mechanisms that governments use to promote the interests of the dominant class and to sustain a particular discourse and policy direction. This analysis highlights the distributional aspects of policy as well as the behavioural aspects such the 'conditionality' of welfare benefits and the policing and self-policing elements of behavioural change. The emphasis on the behavioural aspects of policy provides an insight that is lacking in the other approaches and can shed light on the ways that policies are structured to pursue changes in the behaviour of different agents. The focus on discourse is similar to that of the political approach except that assumptions about the power balance in society differ. Many structural approaches do place emphasis on the balance between agency and structure and assume that policy is not just in one top-down direction. There is a focus on resistance to control mechanisms and on social movements that can impact on the power of dominant discourses and foster alternatives to them. Therefore, Durnova et al. (2016, p. 47) argue that 'the fundamental goal of critical policy analysis is the assessment of standard techno-empirical policy findings against higher-level norms and values'.

This book adopts the discursive approach to policy-making in which the focus is on the language games, coalition-building strategies and power activities that shape policy outcomes. The processes involved in playing out the games and the outcomes achieved are reified into social practices, that is established ways of thinking about and doing things, and become embedded in institutional structures that themselves influence the processes and outcomes achieved. This will be discussed more fully in the following chapter, Chapter 3. It is important to note that empirical studies and evidence can help shape these games, and so the information provided in this book is intended as a contribution to the debates and negotiations around housing policy.

What kind of evidence? Evaluating housing policy

The aim of this book is to examine different forms or mechanisms of housing policy and comment on their effectiveness in different contexts. So, a fundamental issue here is how to evaluate policy. In other words, what does a good policy look like? The existence of different political ideologies and the discussion of policy-making in the previous section highlight that there will be differing answers to this important question.

In some neoliberal thought, there is no need for the evaluation of outcomes as these are determined by competition and market processes. The appropriate role of the state is to ensure that the market functions in an efficient and appropriate way (see Chapter 4) rather than be concerned with outcomes. As Davies (2017) argues, neoliberalism may involve the elevation of market-based principles and techniques to the level of state endorsed norms. There is some disagreement among neoliberal writers about the virtues of competition as the cornerstone of these market principles. In some readings of the neoliberal ideology, markets are held to function most effectively where there is competition between producers and between consumers that is said to lead to maximum utility as well as the moral virtue of liberty. Efficiency of the market may be enhanced through the establishment and enforcement of rules of competition or the reduction in transaction costs associated with buying and selling the commodity. However, some writers (see Davies, 2017, for a review) are more relaxed about competition and are prepared to accept the market forms that emerge from competition, even if they constitute monopoly or oligopoly, as they argue that the market will find the most appropriate form for the particular commodity and context. They argue that monopoly can emerge because, in a particular context, monopoly providers have a competitive advantage, and so this form will lead to the greatest utility. Therefore, advocates of this approach advise government to refrain from intervening in markets.

As well as competition, there are other criteria that may be used to evaluate market processes. One is the concept of *elasticity*. Markets can be judged on how sensitive they are to changes in supply and demand and we will examine this in Chapter 4. For example, is the supply of housing in a particular market sensitive to changes in demand? A second concept is the *transaction costs* of buying and selling a product. The lower the transaction costs to both sellers and buyers, the more transactions there are likely to be in the market, thus aiding the sensitivity to consumer demand. Again, this will be examined in Chapter 4.

However, most political ideologies focus on housing outcomes, and even neoliberal governments have, in practice, been concerned with the impact of policies, and so this is the focus here. In the rational approach, a good policy can be said to be one that achieves the objectives laid down for it. But, as we argued above, objectives may not be explicitly stated, and not everyone may agree what they are, either in general or for specific policies. The political approach to policy-making alerts us also to the existence of different sets of objectives held by groups in society with varying power resources. The objectives may be set by powerful elites and may not reflect the values or perceptions of less powerful groups. Rather than accepting the dominant discourse and the objectives that flow from it, analysis should take into account the different perspectives and values that are implicit in alternative discourses as the discursive approach to policy-making emphasises. Many policies have unintended outcomes as well as intended ones and these may be important in terms of their acceptability to politicians and people in general, as well as the holistic judgement of their worth. Many of the most important housing objectives relate to the condition, distribution and affordability of housing, and so measuring the success of policies depends on being able to identify and assess their impact on the housing outcomes of individual households. Therefore, evaluation has to focus on outcomes. However, this is not as simple as it sounds. There is a debate in the policy-making literature between outputs that are the direct result of policy or the outcomes that are the impact on the lives of people and on the society in general. For example, the output of a policy to support new house building may be considered to be the number of houses built, but the outcomes are the feeling of security and well-being felt by the people who make homes in the houses and the impact on the well-being of the wider society, for example, through greater social cohesion.

The impact of housing policies may differ between individuals and groups. For example, a policy may provide financial support to housebuilders and increase their profits, but this policy may not result in more houses being built and so may not help those who are looking to buy a house. Some policies may lead to an increase in the price of existing houses, which helps those already living there, but may hinder those seeking to buy. Differences in income, wealth, gender, ethnicity and physical and mental ability may result in different outcomes from the same housing policy. Therefore, evaluation should ensure that the variable impact is included.

In addition, the impact of housing policy on households may take a period of time to work through. For example, policies that hinder access to owner-occupation when people are young may have an impact when households reach old age in that they may not be able to generate the housing wealth to use for the provision of support, if it is needed when a person is older. The concept of housing pathways (Clapham, 2005) highlights the important routes through housing situations that households take over time and how decisions taken at one time may pattern the opportunities and constraints of future action. The important lesson when evaluating housing policies is to understand that impacts may take a long time to be evident through the housing pathways of households.

So, given these problems, how can the impact of housing policies be assessed? There is no right answer to this question, and there may be disagreement reflecting different political and value positions as we discuss later. However, the focus in this book is on the long-term impact of housing policies on the individuals concerned, in other words, those who live in or potentially could live in the houses, who may be termed dwellers. The benefits to other parties such as landlords or developers are judged purely on the basis of the benefits that accrue to dwellers. The justification for this position is a value judgement that the sole objective of housing policy is to improve the situation of dwellers and not, for example, to enable developers and landlords to make profits. Profits may be necessary in some housing systems to provide the context for housing outcomes for dwellers to improve. However, it is argued that the outcome should be the sole assessment criterion rather than the profitability of a developer, which is a means to achieve an end rather than an end in itself.

In assessing the outcomes of policies for dwellers, a key concept is that of well-being (see Clapham, 2010; Clapham et al., 2017). Well-being is a subjective concept that is concerned with the feelings of individuals. This brings a recognition that individuals will differ in their responses to life circumstances, although most people report themselves to be happy most of the time even if they are living in a dreadful situation. There is a recognition in the subjective nature of the concept that one person may be happy in a housing situation that may make another person unhappy. The primary aim is to measure the underlying differences in subjective feeling between individuals and over time. Much of the empirical literature, particularly in economics, has adopted life satisfaction as a single metric of well-being. Life satisfaction is typically approximated by asking an individual how satisfied they are with their life on a numerical scale. The justification for the focus on life satisfaction measures is that they offer a cost-effective means of capturing a meaningful portion of well-being. The main weakness of life satisfaction indicators is that they only capture a part of well-being. Someone can be satisfied with their life but still be reasonably judged to have low well-being. A husband who cares for his terminally ill wife may be thoroughly depressed and bored, and at the same time feel a sense of meaning and fulfilment, and may therefore be satisfied with his life, but can he be considered to have high well-being? While life satisfaction, it could be reasonably argued, is a component of well-being in its own right, few would argue that it can be directly equated to well-being.

The theory of subjective well-being addresses this weakness by supplementing life satisfaction with two other components of well-being: high frequencies of positive affect (joy, elation, contentment, pride, affection, ecstasy) and low frequencies of negative affect (guilt, shame, anxiety, stress, sadness, depression). This approach has been notably adopted by the OECD (OECD, 2017). While these three components are correlated, they are also distinct, and so well-being is not one continuum but three, and by only looking at one component, we only gain a partial understanding of well-being. Well-being is an inherently ambiguous concept in that every individual cannot be precisely ranked in terms of their well-being. Nor can it consistently be said whether an individual has higher well-being in one state than another. This ambiguity should not, however, impede an examination of the determinants of well-being. Indeed, the empirical evidence suggests that different definitions of well-being are either moderately or strongly correlated, so an individual who has high well-being according to one definition is also likely to be high according to another definition.

The difficulties in measuring well-being should not detract from its usefulness as a yardstick of housing policy. It is possible to assess changes in well-being over time both for individuals and for the society as a whole and to link these to housing policies (see, for example, Foye et al., 2017). However, the concept of well-being has a number of drawbacks which mean that it should be used with other measures. Three important ones are discussed here.

The first drawback is the difficulty faced by all utilitarian approaches in dealing with trade-offs in utility between individuals and groups. In other words, what happens when a particular policy results in an increase in one person's well-being and a decrease in another's? Utilitarians have had recourse to the Pareto criterion with its assumption of potential recompense by the winner to the loser even though this did not have to happen in practice. However, this criterion seems to be inadequate in a public policy situation where losers and winners could have differences in resources and power. Deciding between the competing claims of different social groups is the essence of politics and should be decided through political processes and guided by political philosophies of social justice in addition to questions of subjective well-being.

The second drawback is the influence of status considerations on well-being. Some goods, including housing, are, at least in part, positional goods in that the satisfaction they provide is related to the position of others. In other words, the goods can confer status benefits (and losses) on individuals. For example, in housing, Foye et al. (2017) have shown that the benefits of increased well-being enjoyed by new owner-occupiers are offset by the reduced benefits of existing owners who see their preferential status diluted and by remaining renters who see their marginal situation increased. As in the distributional question, issues of status difference need to be guided by political philosophies.

The third drawback is the reliance on the subjective judgement of individuals that is, at least partly, influenced by the context. People adapt to their situation and judge their life satisfaction and happiness on the basis of their expectations and experience. For example, Clapham et al. (2017) have shown that housing satisfaction increases initially with a move to a larger house but returns to the previous level after a few years. Is it acceptable that poor people are happy despite their difficult circumstances? If a homeless person reports a high level of well-being living on the streets, should they be left there? The existence of this process of adaptation raises moral issues that will be considered later, but it also leads to a questioning of the status of subjective judgements. Should they be accepted unquestioningly, or is there a place for an outside judgement on what is acceptable? Despite the problems, it is important to retain a focus on subjective judgements, because to question them undermines the agency of the individual concerned. To disregard their view of their situation is to recourse to paternalism or authoritarianism. The focus on subjective well-being is a good start to any evaluation because of the centrality given to the attitudes and perceptions of the people concerned, but there may be a need for an additional moral or ethical contribution.

Therefore, the assessment of policy outcomes such as through the concept of well-being should be assessed against higher-level norms and values. A useful way of integrating the moral and political judgements into concepts of well-being is through the capabilities approach (Sen 1985; Nussbaum, 2011), which is highly attractive because it accounts for the plurality of goals – equality, liberty, as well as well-being – that society may seek to maximise. The capability approach makes a distinction between people's *functionings* (the actions or experiences that people have reason to value) and their *capabilities* (their effective opportunities to achieve these functionings). It is generally thought that policy should focus on enhancing people's capabilities, so as to provide a range of possible ways of life and avoid imposing a particular notion of the good life on an individual level. Therefore, it could be considered more important that public policy maximises the *freedoms to achieve well-being* rather than maximising well-being per se. To give just one example, providing a homeless person the option of shelter at night represents progress (or a more just state of affairs) even if the homeless person turns down the offer, because their *capabilities* – or substantive freedoms – have still been enhanced. The capabilities approach also implies that we should define equality not in terms of resources but in terms of capabilities. A person who is wheelchair bound is likely to require greater resources to achieve a healthy living environment than a person with no disabilities. Just ensuring that there is a threshold of resources available for all does not take into account other factors that enable or constrain what humans are actually capable of with these resources. In a just society, housing policy should account for this range of 'conversion factors' by offering more resources to those people for whom capabilities, like a healthy living environment, are more difficult to achieve. Nevertheless, many empirical applications of the capability approach have generally focused on functionings as these are easier to measure and more available.

An important intermediate concept in applying the capabilities approach to housing is that of 'affordance'. Clapham (2011, 2017) has used the concept to explore the activities and meanings that houses make available to individuals. An important element of the concept is that it includes both use (for example, activities of daily life) and meaning elements (such as a feeling of control or a positive self-identity) and so takes into account subjective feelings and practical activities that it sees as being intrinsically bound together. A fuller discussion of the affordances approach is given in Chapter 5. The concept of affordances is crucial to the application of the capabilities approach to housing, because it provides the framework for identifying the capabilities that a house enables. The key question to be answered is what affordances does a person's housing situation enable them to enjoy.

Despite the growth of interest in the capabilities approach as a way of structuring social science and policy analysis, there is relatively little substantial research that applies the capabilities approach to housing. Therefore, it is important to identify the capabilities that are appropriate for housing, and there have been two approaches to identifying the capabilities that should be taken into account. Nussbaum (2011) has identified a list of those capabilities that she argues are 'basic', and so every country should include them. In her list and others built on it, 'adequate shelter', 'housing' and 'control over one's space' or 'enjoyment of home' appear as core components of capability lists.

Although Sen explicitly refrains from drawing up a list of capabilities (Sen, 2010), he implicitly recognises that there are *basic capabilities* – being healthy, well-nourished and educated – which should be universal. The first priority for any government should be to meet those basic needs – to expand the substantive freedoms that individuals have to be nourished, educated and healthy. To this end, housing policy could ensure that every individual has the option of a home that is healthy. It could be made a legal requirement that tenants can freely demand landlords remove damp, condensation from their home and remedy other housing conditions which have been found to be detrimental to health. Housing policy could also be used to facilitate education and nourishment through ensuring that individuals have the option of a home that is sufficiently sized that children can study in private and adequately provisioned that people can cook. Furthermore, housing should be sufficiently affordable that individuals can meet these basic housing requirements without jeopardising their health, nourishment and education through other life domains. If, for instance, an individual has to forego medical insurance to be able to afford a basic but healthy home, then addressing this clash in basic capabilities should be a key priority for housing policy and public policy more generally. For example, recent cuts to housing benefit in the UK, which have been detrimental to the mental health of private renters, would appear represent a backward step in the pursuit of justice.

However, the factors considered in this 'basic' approach may be small, and some governments may wish to move beyond these. Sen (2010) argues that the capabilities chosen are relative to the context and political choices in each individual country. Therefore, he argues that they should be determined through a process of deliberation and debate through the democratic processes within each country. Therefore, if we want to move beyond the identification of basic affordances, it is likely that people adhering to different political philosophies would compile different lists, and one way forward would be to do this for the different welfare regimes identified. Therefore, in the next chapter, the attempt is made to identify the important capabilities that would be commensurate with the political philosophy inherent in the different welfare and housing regimes. One of the key differences between regimes is their attitude towards inequality. It can be argued that financial and other resources are crucial influences on the capabilities or affordances that individuals receive from their housing situation because of its influence on their ability to compete in the housing market. Therefore, inequality between different groups in a society will be given some prominence in the analyses in this book, although it is recognised that different political choices will be made about the desirable level of inequality in any society in general or housing regime in particular. Nevertheless, a key argument of the book is that inequality is at the heart of many housing problems that will not be solved unless existing levels of inequality in many societies are reduced.

The concept of sustainability that is explored in Chapter 11 adds an additional dimension to the use of subjective well-being as an evaluative criterion. As we argued above, well-being is essentially a utilitarian concept that lacks a moral and ethical dimension. Also, like all utilitarian concepts it is limited in that it lacks time and collective dimensions. Sustainability focuses attention on the impact of policies on future generations, and this may not be picked up in well-being measures. A focus on time highlights the collective dimension, as dwellers are being asked to take measures that may inconvenience or have cost implications for them as individuals for the good of their children and young family members and future generations as a whole. The importance of sustainability is outlined in Chapter 11, and it is relevant to many housing processes and outcomes. Therefore, in this book, it is considered as one of the ethical issues (together with inequality) that should be applied to evaluations of housing policy.

Conclusion

In this chapter, we have put forward a definition of housing policy as 'any action taken by any government or government agency to influence the processes or outcomes of housing' and discussed the issues that this definition raises in terms of the prime role given to government at different levels in taking responsibility for housing outcomes. We have then examined the tools or mechanisms available to government to achieve their housing policy objectives such as direct provision or regulation. Further, the chapter has considered the question of how housing policy is made and the role of evidence, such as that provided in this book, in policy formation. A discursive approach to policy-making has been adopted with an emphasis on the language games, coalition-building strategies and power activities that shape policy outcomes. Empirical studies and evidence can help shape these games, and so the information provided in the book is intended as a contribution to the debates and negotiations around housing policy rather than as a definitive answer to housing problems. Nevertheless, a discussion of the language games, coalition-building strategies and power activities allows us to assess the potential for adopting different policy ideas in specific contexts.

Finally, the chapter has focused on the type of information that is necessary to provide evidence for the policy-making process on what housing policies are effective and what are not. It has been argued that the key group to focus on are 'dwellers' – those who are or could be housed – and that housing policies should be judged on their impact on them. The concept of subjective well-being has been put forward as the basis of a mixed approach that focuses on the capabilities or affordances of houses and neighbourhoods for their occupants. Therefore, a good housing policy is one that improves individual and collective well-being through increasing the affordances that dwellers achieve from their housing, and this yardstick will be used in the rest of the book to evaluate the outcomes of different housing policies where evidence exists. A theme of the book, that will become more evident as it proceeds, is that the necessary evidence to asses many policies does not exist, and so the book is also a plea for more research directed to this end.

Finally, the importance of values and political ideologies in judging the impact of housing policies has been highlighted. Very often a particular policy will increase the well-being of one group of dwellers at the expense of another, and so questions of social justice arise that are contested in political debate. The two concepts of inequality and sustainability have been put forward as two important values by which housing policy should be judged. In this book, this is partly dealt with by reference to the political ideologies inherent in different housing regimes, and this is the focus of the next chapter.

Chapter 3

Housing regimes

The aim of this book is to examine housing policies in a number of countries, with a view to judging their success in achieving well-being for dwellers and to see what can be learned from experience in different contexts. This chapter focuses on the international dimension of the analysis. In the previous chapter, the discursive approach to policy-making was adopted in which the focus is on the language games, coalition-building strategies and power activities that shape policy outcomes and are reified into a set of social practices. In any given country, these practices will constitute a housing regime. The concept was introduced in Chapter 1 and defined as 'the set of discourses and social, economic and political practices that influence the provision, allocation, consumption and housing outcomes in a given country'. The concept of a housing regime allows us to examine both the agency of different actors and the structural elements of housing policy.

The book examines some examples of housing policy mechanisms in a number of countries, but it is important to note that the book does not aim to provide a systematic international comparative analysis. Rather, experiences in six different countries (which will be introduced at the end of the chapter) are offered as examples in order to make particular points about housing policy. Clearly there are limits to what can be learned from looking at a number of countries if their differences are marked and if the context within which housing policy is pursued varies considerably. The importance of the context of housing policy is emphasised in the literature on policy transfer (see, for example, Dolowitz et al., 2000). There are many examples of policies and practices being transferred from one country to another with very poor results. For example, a number of attempts were made to import Scandinavian-style housing co-operatives to Britain with very little success because of the very different social practices, legal structures and cultures within the countries (see Clapham and Kintrea, 1987). Despite potential problems, looking at examples of housing policy in different countries may serve to open our eyes to new possibilities or help to identify lessons that can improve policy and provision. The important factor is to be aware of possible problems with potential transfer and to be alert to the importance of the context within which particular forms of provision are embedded in a particular country. In order to do this the concept of housing regime is adopted in the book, which enables us to highlight the similarities and differences between countries in the extent and objectives of state intervention. It facilitates a degree of categorisation of countries into similar groupings, which is necessary if we are to be able to generalise experience across countries. Examining examples from different countries allows specific factors to be isolated and their impact gauged. In other words, it enables us to understand how policies have taken the shape they have and had the impact they have in different contexts.

The housing regime approach enables the context in a particular country to be explored in a consistent way and gives a framework for the context of policy to be compared. Therefore, it may be possible to argue that a particular policy mechanism that has a specific set of impacts in one country would have the same or similar impacts in another country with a similar housing regime. Awareness of the impacts in the regime category or setting and the contextual factors involved may also enable a judgement to be made about the impact that the policy would make in another setting or regime. For example, on the basis of an analysis of the impact of the policy of the provision of social housing for low-income households in a number of different housing regimes, we may be able to draw the conclusion that the policy has one set of impacts in a neoliberal regime and another in a social-democratic regime. An analysis of why this is the case may enable us to make a judgement on what the impacts would be in a different housing regime.

In this chapter, we introduce the concept of housing regime by examining a number of approaches to the explanation of the similarities and differences between countries. The first approach to be described and evaluated is the welfare regime approach based on the assumed importance of political ideologies. This is followed by a review of the path dependence approach that emphasises continuity and the importance of the institutional structure within individual countries. The third approach is labelled the 'varieties of residential capitalism' and focuses on the financialisation of housing having major impacts in many countries because of the growing links with the global financial system. This discussion is used to develop the adopted approach to the concept of housing regimes that aims to be holistic and to take into account the factors highlighted by the three approaches.

Welfare regimes

The concept of welfare regimes is based on the work of Esping-Andersen (1990) who examined welfare expenditures in a number of European countries and categorised the resultant differences into three regime types, the social-democratic, the conservative and the liberal (which is equivalent to what we label here as neoliberal), which he argued had 'qualitatively different relationships between state, market and the family' (Esping-Andersen, 1990, p. 26). Zhou and Ronald (2017, p. 255) sum up well the main regime types,

> in social-democratic welfare states, universal social rights are given to a large proportion of the population based on citizenship, with the state dominating provision. The regime is thus highly de-commodified with low stratification. The market is often crowded out and the cost of raising a family is also socialized. In conservative welfare states meanwhile, the distribution of social rights is often based on class and status, consolidating divisions among wage earners. Social policies characteristically maintain social differentiation with individual welfare conditions modified by non-state providers: faith-based communities, trade unions, kinship networks, etc. Conservative regimes thus ensure both a measure of de-commodification and a high level of stratification. It is often shaped by the church and committed to the traditional role of family. Thirdly, in liberal (or neoliberal) welfare states, governments usually ensure limited well-being for the very poor, providing some welfare services based on means testing. The policy regime typically seeks to maximize the function of the market while minimizing the state's involvement. There is thus a low degree of de-commodification and high levels of stratification.

According to Esping-Andersen, a regime constituted the power relationships that structured welfare provision in a country, and he focused particularly on the power of the labour movement in structuring his regime types. This concept of a regime is narrower than the one taken in this book that focuses on both the power structures around the provision of housing and the type of housing itself, as we will discuss in a later section. It is important at this stage to be clear about the nature of the categories used in the welfare regime approach and how they are determined. Esping-Andersen undertook his categorisation on the basis of actual welfare expenditures, and so his categories were clusters of countries based on actual policies. He then generalised across these countries to construct his regime type, although he was aware of differences between countries in the same cluster. Therefore, his generalisations were not 'ideal types' in the Weberian style that would be logically consistent models created from idealised concepts, as Esping-Andersen's came from the grouping of countries based on data. Nevertheless, he used them like ideal types in the sense that the categories he created are abstract creations that are meant to show possibilities that actual practice can be compared with, rather than 'real life' categories themselves that would accurately describe the situation in any individual country. Housing studies have tended to follow Esping-Andersen's approach by constructing ideal types from empirical research on particular countries, but the core point about whether the categories are descriptions of actual situations or logically consistent ideal types is often left unclear.

Esping-Andersen himself saw the political power of the labour movement as being the factor that primarily influenced the regime of any particular country. Others (see, for example, George and Wilding, 1976) have sought to give more depth to the categorisation by exploring the ideological basis of the categories. Esping-Andersen's regime types do relate closely to the major political ideologies that have shaped political discourse in the advanced capitalist countries, although they need to be expanded to include ideologies that are prevalent in other countries outside this narrow core. This analysis places emphasis on the influence of political choice in creating and sustaining a particular housing regime.

It is interesting to note that the original categorisation was based on welfare services and did not include housing, perhaps reflecting housing's situation as the 'wobbly pillar' of the welfare state (Torgerson, 1987). However, the approach has been widely used in housing research with important contributions from Kemeny (1991) and Castles (1998) to add housing into the analysis. Kemeny identified three causes of housing regimes: the balance of power between capital and labour, its mediation through social and political structures and what can be thought of as an underlying ideology. Kemeny's major contribution was to make the link between owner-occupation and other welfare services. He argued that owner-occupation gave households an asset that could be used to fund old age and so made them less likely to vote for universal pension provision. Also, he argued that owner-occupation was an individual form of consumption that made households less likely to support universal and collective forms of welfare and other state services. Castles confirmed the link between high rates of owner-occupation and lower welfare spending in a wide range of developed countries but argued that the causation could be the opposite to that suggested by Kemeny. In other words, low welfare spending would encourage households to enter owner-occupation in order to protect themselves by individualising consumption and building up an asset to provide security. Following Kemeny's and Castle's analysis one would expect countries with neoliberal (or liberal using the original terminology) welfare regimes to have high rates of owner-occupation and social-democratic regimes to have low ones and high rates of state-provided rental housing. The extent of owner-occupation may be linked to ideologies of 'asset-based welfare'. Here the expectation is that people will build up capital during their working lifetime in order to fund their life in older age. The major asset for most households is in an owner-occupied house, and so housing is at the core of this approach. If people can be persuaded to build up equity in their house to finance their retirement, then welfare benefits such as state pensions and services such as health, social care and residential accommodation can be cut back and reserved only for those with no assets. Of course, there needs to be adequate means of unlocking these assets through equity release products or through trading down to smaller and cheaper accommodation. Also, this approach has profound distributional consequences as universal welfare services available to all will be replaced, at least in part, by benefits based on the ownership of capital that is very unevenly distributed between people.

Kemeny also made the distinction between unitary and dual rented sectors. A unitary sector is dominated by the public sector and by state regulation that forces the private sector to play a subsidiary and complementary role. In dual rental markets the public sector is not allowed by government to compete with private rental and so occupies a residual role confined to those unable to be housed in the private sector. The distinctions in tenure structure form the basis for the integration of housing into the welfare regimes. A neoliberal regime is likely to have a high rate of owner-occupation and a dualistic rental system. A social-democratic regime is likely to have low owner-occupation and a unitary rental system. A corporatist regime is likely to have low rates of owner-occupation and low public rental with a large private rental sector run on dualist lines.

Hoekstra (2003) related the characteristics of the welfare regimes to particular aspects of housing systems. For example, he linked the concept of de-commodification to the housing subsidy and price regulation system; stratification to housing allocations; and state, family and market mix to the systems for the production of new housing. Therefore, the nature of housing systems is susceptible to the welfare regimes approach.

One criticism of the early work on welfare regimes was the limited number of countries considered as well as the relative lack of focus on the family element of Esping-Andersen's original trilogy. With a widening of the scope of analysis, a number of different regime types have been added, such as a familial model based on the Southern European rudimentary welfare states (Leibfried, 1992). In housing, these countries tend to have high rates of owner-occupation and small rental sectors, with most building taking place through familial structures and being 'self-built' often on family-owned land with an under-developed speculative development industry and very limited state involvement (Allen, 2004). Another addition has been an East Asian regime based on Confucian values that stress the importance of family and community mutual support (Lee and Ku, 2007). However, in the case of China, which is reviewed later, this has been superseded by a concern with a 'productivist' regime where social (and housing) policy is aimed primarily at economic factors. There was also consideration of an Eastern European or socialist regime during the transition from communism in the 1990s, but it is now generally accepted that most of these countries have moved towards a neoliberal welfare regime (Matznetter and Mundt, 2012). However, the regime type has relevance still in countries with periods of socialist political ideologies such as certain times in China.

Studies have been almost entirely concentrated on the developed economies, and few have considered countries in South America or Africa. This neglect may be due to a realisation that context matters, and the economic, social and demographic context facing countries in these regions is very different from that of the advanced capitalist countries usually studied.

Although the welfare regimes approach does include institutions and, to some extent, the outcomes of welfare ideologies, these are tied in with the political ideologies in the regimes identified, and so it is difficult to isolate their influence and to distinguish between regimes where similar ideologies have been implemented in different ways. The type of institutions may differ according to cultural and historical factors rather than political ideology. Therefore, we need to examine other approaches to be able to construct a clearer picture of the important factors in the housing regime that vary between countries.

Path dependence

An initial criticism of the welfare regimes approach is that the categories were constructed on the basis of circumstances at one point in time. Authors who stress path dependence in housing argue that the regimes are constantly changing and analysis should examine these trajectories over time (see Bengtsson and Ruonavaara, 2010). Path dependence can be defined as 'a historical pattern where a certain outcome can be traced back to a particular set of events on the basis of empirical observation guided by some social theory' (Bengtsson, 2008, p. 5). The argument is that housing policy is usually incremental in nature and is strongly influenced by the institutions involved in an ongoing and self-reinforcing chain of games played by actors and institutions in which change is limited by the power of actors to define what is legitimate, the costs of change and the power vested in existing institutions. There is a close similarity here with the concept of the discursive policy-making outlined in the previous chapter. Path dependence recognises that there are 'critical junctures' in which times a change of direction takes place that brings about more radical change. They argue that 'history matters' if one wants to understand the dynamic nature of housing regimes. A path dependence method is to find a point at which a key housing decision is made by government and then to trace back to find the critical juncture at which the factors that made this decision necessary and possible become apparent and to then to analyse the mechanism by which the two points were related (Bengtsson and Ruonavaara, 2010).

Bengtsson and Ruonavaara (2010, p. 195) argue

> the typical case of path dependence is where actors design institutions or make policy decisions at point (or points) A, which at a later point B set the rules of the political game between the same or other actors. In retrospect, the historical development can be perceived as an ongoing and self-reinforcing chain of games between actors, institutional change, new games, new institutions, etc.

Therefore, authors in this tradition recognise the importance of the existing state of affairs in legitimising current activities and creating boundaries of what is acceptable in policy debate. Again, this has many similarities with the discursive approach outlined in the previous chapter. Bengtsson and Ruonavaara (2010, p. 195) state,

> This means that the (relatively) contingent events at point A would make some alternatives appear either to be more efficient, more legitimate or more powerful at point B. The *efficiency mechanism* of path dependence has to do with the coordinating capacity of established institutions and the transactions costs of changing them. The *legitimacy mechanism* may influence either what political actors themselves see as legitimate or their perceptions of what is legitimate in the society at large. Correspondingly the *power*

mechanism may affect either actors' own power or their perceptions of power relations in the larger society. The power mechanism may also have an impact on which actors are allowed to take part in the decision-making at point B.

A key element of path dependence studies is a periodisation based on the critical junctures when political choices have been made to change housing policy, although sometimes the critical juncture is the housing policy change and sometimes it is the presumed factor that triggered the change. This can sometimes lead to a very mixed categorisation with housing factors (such as a choice to produce large-scale, state-provided rental housing) alongside trigger factors (such as wars or economic crises). However, most studies have focused on the housing policy changes. Following their study of housing policy in the Nordic countries, Bengtsson and Ruonavaara (2010) identified four historical phases based on housing policy:

> an establishment phase with limited housing reforms in response to the early urbanisation; a construction phase with comprehensive and institutionalized housing policies aimed at getting rid of housing shortage; a management phase where the more urgent housing needs had been saturated; and a retrenchment phase with diminishing state engagement in housing provision.
> (Bengtsson and Ruonavaara, 2010, p. 196)

The five Nordic countries were said to have passed through these phases at different times as they were subject to the same structural forces of industrialisation, wartime crises, mass construction and privatisation. However, their housing regimes remained different from each other and reflected the historical forms that provision had taken in each country. In their study of housing policy in the four largest Latin American countries, Murray and Clapham (2015) identified three phases of housing policy: spreading informality and accruing social housing debt; acceptance of the Washington consensus; and building the way out of recession. All of the four countries passed through these phases at different times and with different impacts on their housing regime depending on the nature of their housing institutions. In studies of Eastern European countries, Clapham (1995) identified a set of changes that were occurring in the housing regimes in these countries as they moved towards western economies. Therefore, periodisation based on changes in housing policy is a key feature of studies using the path dependence approach.

Malpass argues that the continuities identified by studies of path dependency in housing are appropriate at a very general level. Therefore, he argues that the general aims of British housing policy have been constant over a long period of time, but that there has been substantial change at a more operational policy level. More fundamentally, Malpass (2011) criticises the basic assumptions about change in the path dependence approach, which he argues are contradictory. The approach assumes that there are long periods of incremental change in one direction determined by the forces of existing institutions, but these are interrupted by critical junctures that result in substantial change. Malpass argues that, in reality, housing regimes are in constant incremental change that can lead to radical departures over a period of time. He uses the example of the switch from local councils to housing associations as the major provider of social housing in Britain to show that a series of small changes, often not with any particular end state in mind and sometimes accidental, can have significant results. Therefore, he argues that the path dependence focus on critical junctures is a misleading one, as is the assumption that change is not an ongoing phenomenon. He also takes issue with the assumption that moves down any particular path make a reversal of direction less likely. He argues that there is an ongoing process of review in which housing policies are judged on their impact and can reach a point at which they can be shown not to 'work' or to be ill-suited to a new situation and so be changed. In reality, he argues that there is not incremental change in one direction with occasional radical moments, but constant minor change in many different directions at the same time. He cites Streeck and Thelen (2005, p. 31) who introduce a more complex picture of change by identifying a number of ways that the institutional dynamic that results in the unidirectional and incremental change can be altered:

- *Displacement*: slowly rising salience of subordinate relative to dominant institutions;
- *Layering*: new elements attached to existing institutions gradually change their status and structure;
- *Drift*: neglect of institutional maintenance in spite of external change resulting in slippage in institutional practice on the ground;
- *Conversion*: redeployment of old institutions to new purposes; new purposes attached to old structures; and
- *Exhaustion*: gradual breakdown (withering away) of institutions over time.

Malpass identifies the existence of both displacement and layering in his study of the move from local authorities to housing associations in Britain. Nielsen (2010) uses the concept of institutional weariness to understand change in Danish housing policy. She identifies three hypotheses of change in housing policy:

1 Change can happen either as intended action, non-decision-making or unintended consequences.
2 Housing policies change more often through *drift, conversion or layering* than through high-profile reforms due to the high level of institutional feedback-mechanisms and status quo in the general environment.
3 Information on the consequences play a key role – as more information on the consequences of the changes reach the public the status quo pressure on the policy-makers increases, making radical change less likely.

The path dependency studies have been at pains to point out the differences between the housing regimes in the countries studied, as one criticism of the welfare regimes approach has been the neglect or de-emphasising of these differences. However, this criticism is less applicable to categories used as ideal types, where the aim is just to highlight issues through comparison with an ideal. Nevertheless, the 'path dependency' approach does highlight the importance of housing institutions (including both organisations and social institutions or social practices in housing) in influencing the shape of the housing regime. Similar trends may take a unique form in different countries because of the objectives, priorities and forms of the different institutions involved. For example, owner-occupation may mean different things in terms of the duties and rights of owners and the meaning of the tenure may vary in different circumstances (see Mandic and Clapham, 1996). Rented housing may be produced and managed by local authorities or housing associations. Social structures may also be important. By this is meant the shared meanings that may exist in a country about the status of different tenures or the way that housing is consumed. Institutions are viewed here as sets of regularised practices with a rule-like quality in the sense that the actors expect the practices to be observed and which, in some but not all, cases are supported by formal sanctions. They can range from regulations backed by the force of law or organisational procedure, to more informal practices that have a conventional character. An important element of the social practices in a housing regime is made up of the mechanisms used by government to achieve its aims. A range of these including regulation, direct provision and the taxation and subsidy mechanisms was highlighted in the previous chapter. As is argued in Chapter 4, governments make the housing market and the housing regime, and the mechanisms they use to do this, are themselves major elements of the regime.

Varieties of residential capitalism

In the discussion so far, emphasis has been on the factors within a country that can influence the shape and trajectory of the housing regime. However, it is argued that there are some general factors that may shape all regimes to some extent. For example, it has been argued that financial globalisation is one of these factors (Clapham, 2006). Kemeny and Lowe (1998) drew attention to what they labelled as 'convergence' approaches to comparative housing that highlighted the similarities in the trajectories of different countries in the face of global influences. In particular, it is often argued, compellingly, that most countries are moving towards a more neoliberal and free market approach. Whilst not accepting the 'convergence' argument, the 'varieties of capitalism' approach does emphasise the impact of global forces and, in particular, places the focus on the housing finance structures and how they interact with global financial markets. It is argued that financial deregulation from the 1990s onwards allowed for the financialisation of global capital, meaning that 'profit-making increasingly occurs through financial channels rather than through trade and commodity production' (Aalbers, 2008, p. 148). As a new way of capital accumulation, national housing finance systems became more strongly integrated in the global economy, which in turn became increasingly dependent on the performance of housing markets. Mortgage and other finance institutions came to rely more strongly on market-oriented funds and inter-bank lending (as opposed to their deposit base) to finance domestic activities. One form of this was mortgage securitisation that meant that mortgage debts were packaged together and sold on the world financial markets and which, when they became increasingly problematic because of a growth in mortgage defaults, became a major factor in the Global Financial Crisis (GFC). However, countries varied in the degree to which they deregulated their housing finance markets, and some (such as Germany, Austria and France) maintained controls that limited the exposure to global finance markets, and so they avoided some of the major effects of the GFC.

Aalbers (2015) identifies a periodisation of housing policy by documenting a shift from what he terms a 'pre-modern' period to a 'modern/Fordist' era of housing policy to a 'flexible neoliberal/post-Fordist' one that laid the conditions for the Global Financial Crisis in 2008 that led to the present 'emerging post crisis/late neoliberal' era. The 'modern' era was characterised by large-scale housebuilding often through the provision of social housing, whereas the developments in the neoliberal era were the neglect of social housing, the growth of owner-occupation and the financialisation of housing. He argues that the period leading up to the 2008 Global Financial Crisis was characterised by a growing synchronisation of housing policy across many countries in both the developed and developing world. He argues that:

> One effect of globalisation is that contemporary countries go through the same developments at roughly the same time. This does, of course, not imply that globalisation has the same effects around the globe. On the contrary, globalisation is a process of uneven development at the global scale and on a global scale.
> (Aalbers, 2015, p. 46)

The credit associated with commodified housing becomes an increasing part of national economies as well as personal wealth and credit. The emerging post-crisis era is characterised by a reduction in house building in general and in the proportion of people in owner-occupation, as problems of affordability become pronounced. This has caused an increasing problem for younger people wishing to enter owner-occupation for the first time.

Aalbers notes that the GFC has had a different impact in individual countries because of uneven development, brought about by varying exposures to global financial markets. Schwartz and Seabrooke (2009) offer a collection of the experience in a number of countries following the GFC and show the differences on the make-up of the form of residential capitalism due to the economic structure and the openness to global financial markets, as well as the response of governments to this. On this basis, they identify categories of residential capitalism as corporatist market, statist-developmentalist, liberal market and familial. One drawback of these studies is that they are limited to what may be called developed countries, and there is little reference to countries in the global south. Aalbers (2015 pp. 45–46) does note in passing that the same recent trends such as the neoliberalism of policy and growing mortgage securitisation are present in many of these countries and some have a housing system that 'has one foot in the pre-modern period and one foot in the post-Fordist period, some of them entirely skipping the modern or Fordist period in housing' (Rolnik, 2013) where many countries have come straight to this point without moving through the modern period. The assumption here is that the same analytical categories are appropriate in the global south, and this would seem to be confirmed in the case of some South American countries (Murray and Clapham, 2015), a point we will return to in the book as we examine housing policy in some of these countries.

Schwartz and Seabrooke (2009) construct a regime typology on the basis of two important factors that influence the different macro-economic impact of housing finance structures. The first factor is the extent of owner-occupation that indicates how 'commodified' and thus open to financial forces the housing system is. The second factor is whether the housing finance system is 'liberal' or 'controlled'. This is measured by the amount of mortgage debt as a proportion of GDP and the degree of mortgage securitisation, which limits state control over the housing finance system and links systems more closely to general global financial markets. Schwartz and Seabrooke (2009) use these two factors to produce a different typology from that common in the welfare regimes approach. Highly commodified and liberal finance regimes tend to fit the traditional 'liberal' (or neoliberal) welfare regime. However, the traditional social-democratic group is split between what Schwartz and Seabrook call a corporatist-market (which includes some social-democratic countries as well as those in the corporatist regime type) and a 'statist-developmental' category that includes some of the traditional social-democratic welfare regimes with a low rate of owner-occupation and a controlled system.

The strength of the 'varieties of capitalism' approach is that it includes in the analysis the national political economies that condition economic performance and social well-being. This is an important adjunct rather than alternative to the focus on welfare policy, as housing policy and housing outcomes are influenced by both welfare and finance issues and at local, national and global scales. As Aalbers (2015, p. 46) notes,

> housing is not only national in nature, but also local and global. As is well understood, housing is local in nature because housing markets work locally but, in the majority of countries, most housing market institutions and the lion's share of housing policies are embedded at the national scale. Housing is also global in nature because, first, some agents of housing markets work globally, and, second, the ideology of housing as well as of states and markets is shaped in a complex fashion at the intersection of national and international trajectories.

He argues that the there are some common trends in the trajectories of countries, particularly in more recent times, but maintains that this does not imply that there is convergence, as countries still have their distinctive institutions and structures at the local and national levels.

The periodisation suggested by Aalbers is similar to that provided by path dependence theorists. Although its precise form and labelling is more linked to economic factors, the housing policy periods and their defining characteristics are similar. Also, following the varieties of capitalism analysis, it is clear that the Global Financial Crisis in 2008 is a 'critical juncture' that has had profound impacts on many housing regimes. The precise impact though will depend on the nature of that regime. Therefore, a useful way forward seems to be to adopt a periodisation, whilst accepting the existence of different regimes that reflect political attitudes towards welfare and to the relationship between housing and economic and financial structures.

Housing outcomes

Most of the studies of comparative housing reviewed above focus on housing policy and policy-making, and there is a relative neglect of the housing circumstances facing dwellers. Housing circumstances are both an outcome and an input into housing policy-making. The situation faced by dwellers will be an influence on the political process and could help to frame and define housing problems. At the same time, housing policy is predominantly aimed at altering these outcomes in a desired direction (some housing policies are aimed at other objectives as we will see later). This does not mean that the state is seen as a neutral arbiter reacting to housing conditions to create a 'better' housing system, as it is recognised that many other factors influence policy. But, housing outcomes can be used as a criterion for judging the efficacy of housing policies as we showed in the previous chapter in our discussion of the concept of well-being.

Comparative studies of housing have tended to focus on tenure as the defining issue in housing outcomes as well as some aspects of physical housing conditions such as overcrowding or houses in poor repair. Perhaps the most important index of this type relates to the amount that households pay for their housing and its relationship to their income and wealth. Housing regimes may have variable impacts on different income groups with some being redistributive in leading towards greater equality and others reinforcing or strengthening existing inequalities. Therefore, an index of the relationship between monetary and housing incomes is important in defining the impact of a housing regime.

Indicators of physical housing conditions may be important, but they do not offer any insight into how people experience their housing and what it means to them as they are mediated by cultural and personal factors as we shall see in Chapter 5. Therefore, outcomes in terms of the material structures and conditions need to be supplemented and perhaps even superseded by those that capture the personal and subjective impact on individuals by using measures such as housing satisfaction or subjective well-being as outlined in the previous chapter. The concept of affordances is used in the book to cover both the use and meaning outcomes that a person's house can afford them. In other words, it covers factors such as the activities of daily living that a house enables as well as the identity and self-esteem that it may bestow. It was argued in Chapter 2 that the affordances considered important by a government may differ according to the housing regime and the political philosophies that it contains. Nevertheless, for our purposes here it is possible to identify a number of affordances that are useful in comparing the outcomes of different housing regimes.

Different housing regimes may result in patterns of outcomes that differentiate between segments of the population in terms of house conditions, affordability and status and meaning. The latter two factors are particularly important as we noted in the previous chapter because they impact directly on individual well-being. There we argued that the most important outcome of housing policy is the impact it has on the well-being of dwellers, and this is the predominant outcome to be considered.

Housing regimes

In their definition of housing regime, Stephens and Van Steen (2011, p. 1037) draw the distinction between housing regimes and housing systems.

> Regimes, in Esping-Andersen's thesis, are the power structures that give rise to systems that contain certain properties (see also Ronald, 2008). So, in Esping-Andersen's (1990) typology of welfare regimes, social-democratic, corporatist and liberal regimes (power structures) have given rise to welfare systems whose essential characteristics produce distinctive patterns of distribution, notably 'decommodification' and 'stratification'.

The distinction between regimes and systems does not seem to be helpful and ignores the way that the power structures are intrinsically related to the institutions and structures of the housing regime. Therefore, here we use housing regime to describe what Stephens and Van Steen call the housing system as well as the power structures and institutional patterns that are associated with it. Therefore, a housing regime is here defined as 'the set of discourses and social, economic and political practices that influence the provision, allocation, consumption and housing outcomes in a given country'.

The three main approaches to comparative housing analysis have largely been treated as alternatives with analysis following one or other tradition. However, housing regimes are complex and are influenced by a wide range of factors. The three approaches are important in that they highlight some of these influences whilst ignoring others. The approach taken in this book is to bring together the factors highlighted by the different approaches as housing policy is influenced by the policy-making discursive games that are reified in the institutional structure, as well as the welfare ideology and the finance system. Therefore, regime types are established here from a mix of political ideology; institutional housing structures; and social, financial and economic structures. At the same time, it is accepted that the periodisation employed by Aalbers and others gives a useful way of examining the trajectories of housing regimes over time. However, it must be stressed that these are ideal types and do not describe the housing situation or housing policy in the individual countries from which examples are taken in the book.

The different approaches reviewed above all offer insight into the shape of housing policy in a particular country and the factors that influence this. Path dependence is based on the idea that history matters and that the current situation is shaped by decisions made in the past. It uses the concept of critical junctures to highlight times when change is pronounced and important policy choices are made. Although much research in the welfare regimes approach has been criticised for being too static in its analysis, some studies have used periodisations to show how policy can change over time (Murray and Clapham, 2015). However, most insight can be gained from the varieties of capitalism tradition, with its emphasis on the changing world economy and the uneven development of capitalism. Aalbers' periodisation outlined above (Aalbers, 2015) is a good starting point for any analysis. Together with Schwartz and Seabrooke (2009), Aalbers identifies the Global Financial Crisis in 2008 (GFC) as a 'critical juncture' that has resulted in substantial change in housing regimes. Therefore, it seems to offer a key reference point in the analysis in this book.

Residential capitalism

The studies highlighted above focused on the financial form of residential capitalism. This is important in understanding the structure and functioning of housing finance and residential development institutions and markets. For example, in Chapter 1, reference was made to rentier capitalism, and this has been associated with a concern in private companies with short-term share value rather than longer-term profitability which will influence the nature and outcomes of the housing market.

But the housing regime may also be influenced by general factors in the economy. An important one to highlight is the shape and extent of employment markets. For example, flexible labour markets that limit job security and high rates of unemployment are likely to have profound impacts on individual households and on general housing outcomes as well as the shape of the housing system as a whole. In a similar way, the extent of inequality in income and wealth will impact on the ability of individual households to be able to afford housing and to maintain it, as well as the quality they can obtain. It may also shape the structure of tenures and the distribution of housing between groups. Inequality and the labour market may be systematically linked to changes in financialisation, but the link needs to be empirically made, and the factors may have independent impacts that are over and above that caused by financialisation. Therefore, in this book we will take this more general view of the economic sphere as well as focusing on financial structures. The concept of residential capitalism allows housing to be linked with other economic and employment factors that clearly impact on housing policy and outcomes such as the extent of general inequality of income and wealth and the nature of employment markets. Factors such as the extent of unemployment, average and minimum income levels and the security of employment influence the nature of housing.

It is important to take into account the specifics of each country studied, but for the purposes of comparison here, we choose five particular measures. They are the size of GDP, the degree of inequality in the society, the extent of labour market flexibility, whether the economy is controlled or liberal and the degree of financialisation in the housing system.

Welfare ideologies

Despite the assumption of common global trends, national governments have the space for housing policy choices. The response of different governments to the Global Financial Crisis and their structuring and controlling of the residential market will depend to some extent on their political ideology. Therefore, the ideal types offered by Esping-Andersen are still relevant to any analysis if complemented with the more recently identified categories. However, each country has its own variant of political direction, and the mix of national and local government influence in housing means that any categorisation is very general and a simplification of the complex real picture.

Institutional structures

The path dependence approach emphasises the social and organisational institutions that structure the housing regime. The argument is that every country has an individual structure that influences housing outcomes. Examples of the kind of institution considered include meanings of owner-occupation as well as organisational forms for the provision of social housing and the tools and mechanisms used by government to achieve its aims and to 'make the market'. Institutions include the categorisation used by Kemeny between unitary and dual rental systems and the involvement of familial and third-sector institutions. The institutional structure in every country is unique; however, for the purposes of comparison of the countries included in the book four major factors are identified: the extent of housing market dominance, the extent of government intervention, whether there is a corporatist provision structure and the type of social practices. This enables a fourfold categorisation to be made between market dominant, familial, corporatist and government dominant.

Housing outcomes

It was argued above that the outcome of a housing regime is an important element in defining its nature. Although there are many problems in assessing the crucial subjective aspects of housing satisfaction and well-being, it is necessary to consider what we can ascertain about outcomes if we are to understand what the housing regime in different countries delivers to dwellers in terms of their housing circumstances. As we argued in the previous chapter, the most important factor is the impact on the well-being of dwellers. Where this information is available it is a vital element in judging outcomes and is usually assessed in terms of residential satisfaction. Where this is not available the other (output) factors considered here are physical conditions; the extent of housing shortage; and the distribution of housing between different social, income and ethnic groups.

All of these elements go together to comprise the housing regime in a particular country. As noted earlier, the concept of housing regime is used in this book as a way of highlighting similarities and differences between countries. It is particularly useful as a way of incorporating the different contexts facing housing policy in individual countries. Although here we are divorcing housing policy from its context in order to be able to identify the general lessons to be learnt, context is brought back in through the regimes approach so that the impact of policy can be assessed in relation to the different situations. It may be that one policy instrument works in one regime but not in another.

The six housing regimes

Examples from six countries are used in this book to illustrate aspects of housing policy and will be referred to in the following chapters. Therefore, it is important here to justify this choice and to give some background to the housing regime in each country within which the examples can be positioned. Of course, one of the criteria used to make the choice has to be knowledge of the countries concerned, and this restricts the available options considerably. So, the countries here are ones that I have some knowledge of and have written on in the past, although the level of my knowledge varies considerably between the countries.

The six countries are the UK, the USA, Australia, Sweden, Argentina and China. Together these countries offer a wide range of experience in terms of economic development, political ideology and housing circumstances. Nevertheless, these countries are in no way representative of all countries, and no claims are made for this. Given the focus in this book on the impacts of neoliberal housing policies, three countries were chosen with neoliberal political and welfare ideologies associated with neoliberal housing regimes, and this explains the predominance of the so called Anglo-Saxon countries of the UK, the USA and Australia. Sweden was chosen as an exemplar of a social-democratic political ideology, although as we shall see, this categorisation may not be too accurate now, and its housing regime has become more neoliberal over the past decades, although it retains elements of its social-democratic roots. Argentina provides an example of a developing country with a very different economic, social and political context to the others here, and China offers a very contrasting housing regime and political approach.

In the neoliberal housing regimes, there is a predominant housing policy that is based on a neoliberal view of housing markets in which state intervention is kept to a minimum and outcomes reflect the inequalities in the society. Housing policy interventions here are likely to involve regulating and steering markets and encouragement for them to be more efficient. There may be scope for privatisation in the housing system and enlarging the space for profit-making from housing through mechanisms such as privatisation, especially in the UK where there has been substantial state involvement in the past.

The predominance of countries with neoliberal housing regimes reflects the aim of the book in elucidating the nature and evaluating the outcomes of this regime type. Even some of the other countries chosen, such as Sweden and Argentina, show substantial elements of the neoliberal approach and are moving strongly in this direction. China has had periods of neoliberal dominance although it has moved away from this recently because of productivist concerns. The absence of a clear alternative to the neoliberal regime means that it is not possible to compare neoliberalism with another approach. Therefore, the construction of an alternative approach is left until the final chapter.

Although at one time Sweden could be argued to be close to the social-democratic ideal, housing and social policies since the economic problems of the 1990s have resulted in a more market-oriented system. Also, income inequality has increased substantially in Sweden, which was once one of the most equal countries in the world in this measure, but is now becoming more unequal faster than any other country, although it has levels of inequality well below that of the neoliberal regimes described here (Clapham, 2017).

Argentina is an example of the Latin American housing regime type identified by Murray and Clapham (2015) and shares many policy phases with other countries in this region. In the first phase during the 1960s and early 1970s, housing policies in Argentina were fashioned to meet the demand of those in formal employment rather than the unemployed and those working in informal or casual jobs. This was despite the fact that the informal sector counted for more than 60 per cent of the population. The second phase during the 1990s involved the adoption of economic recommendations promoted by the World Bank and known as the Washington Consensus based on a neoliberal ideology. Broadly speaking this implied the reduction of the state, decentralisation, privatisation, opening of markets and deregulation. However, economic problems around the millennium led to attempts to boost housing production through direct state intervention, largely through financial mechanisms, in order to help to combat the economic recession. Despite having governments of the left for a substantial part of the last few decades, the Argentinian housing regime is largely neoliberal with a very high rate of owner-occupation and little direct public housing provision. Housing outcomes are highly stratified by income, and Argentina has a level of inequality equal to the USA. It also has a high level of GDP when compared to other South American countries. Nevertheless, the general economic situation and the residential market are both substantially different from the Anglo-Saxon countries and a European country such as Sweden.

China has been categorised as having a productivist welfare regime that deserves some discussion as it has not been introduced so far. It is suggested (Zhou and Ronald, 2017) that China had a socialist regime from 1949 to 1977 but made a gradual transition to a more neoliberal system up to 1997 based on decentralisation to work units which were marketised and liberalised and engaged in support for owner-occupation. Therefore, housing outcomes were stratified on the basis of an individual's occupation and the financial status and stability of the work unit. From 1998 and up to 2008 the regime could be classified as neoliberal, with the state not only reducing subsidies on low-profit housing, but also loosening its role in price setting. This left house prices subject to market processes, which, in the emerging economic landscape of the 2000s, took a primary role in financing, constructing and allocating homes. In this context, the rental sector withered, and owner-occupied house prices inflated rapidly, stimulating extreme housing stratification. Also, this market system created a high level of migration into the cities, which was difficult to deal with through the market. Therefore, in 2008 housing was announced to be a human right, and programmes of public sector house building were instituted. At first glance this may look like a move towards a social-democratic welfare regime, but differences between cities show that it is directed at the economy rather than social policy objectives and so has been called a productivist regime. In productivist welfare states, which are also advanced economies, economic objectives largely define social policies, with the application of social security as a means to target economically important interest groups. Priority in policy formulation is thus given to enhancing economic and social development, with state, family and market relationships aligned around these objectives. However, the measures needed may vary between countries and cities and regions as the needs of the economy may vary, and there are alternative ways of achieving economic aims. Therefore, commodification and stratification of housing may be high or low depending on the context.

Table 3.1 The six housing regimes

Country	Variety of capitalism	Welfare regime	Institutional structure	Housing outcomes
UK	Liberal/high GDP/flexible/financialised/unequal	Neoliberal	Market dominant	Unequal/good conditions
Sweden	Controlled/high GDP/inflexible/non-financialised/equal	Social-democratic	Corporatist	Equal/good conditions
China	Controlled/medium GDP/inflexible/non-financialised/equal	Productivist	Government dominant	Stratified/improving conditions
USA	Liberal/flexible/high GDP/financialised/unequal	Neoliberal	Market dominant	Highly unequal/good conditions
Australia	Liberal/flexible/high GDP/financialised/medium inequality	Neoliberal	Market dominant	Unequal/good conditions
Argentina	Controlled/inflexible/low GDP/low financialised/highly unequal	Latin American	Market dominant	Highly unequal/poor conditions

Table 3.1 gives a summary of the main factors identified for the six countries used to derive the case studies used in the book.

Conclusion

The previous chapter described the agency elements of policy-making in housing and, in particular, focused on the discursive approach with an emphasis on the language games, coalition-building strategies and power activities that shape policy outcomes. This chapter has put this into context by examining the structural factors that frame policy-making and which are reified into national housing regimes defined as 'the set of discourses and social, economic and political practices that influence the provision, allocation, consumption and housing outcomes in a given country'. The concept of a housing regime allows us to examine both the agency of different actors and the structural elements of housing policy.

This book uses examples from six different countries to explore housing policy, and this chapter discusses some of the opportunities and problems that this focus entails. The concept of housing regime has been introduced and described in the chapter to aid in the international focus and is used in the book to structure the contents of the following chapters. Every housing regime is unique, but following the welfare regimes approach and usual practice in housing studies, it has been possible to identify some key indicators that can be used to categorise and compare the housing regimes in different countries. This is used together with a periodisation based on the path dependence and varieties of residential capitalism approaches that focuses on changes over time. Housing regimes will be used as a concept to aid contextualisation of the discussion of the impacts of housing policies in different countries. Particular housing policies may be successful in some housing regimes and not in others because of the important contextual factors involved.

The first three chapters have introduced the concepts that are used in the following chapters to examine different areas of housing policy, bearing in mind the regime differences that provide the context for the discussion.

Chapter 4

Making the market

In the previous chapter, the concept of a housing regime was introduced that placed emphasis on the institutions, ideologies, economic processes and outcomes that constitute the unique housing circumstances in individual countries. It was stated in Chapter 1 that the housing market is seen as the primary tool for achieving housing outcomes, and so its form and functioning will be a major part of a housing regime. Markets do not necessarily assume a generic form, and governments have multiple and important choices to make over the extent of market coverage and the type of market that exists and the way that it functions. The extent of the market can be influenced by the existence of direct state intervention through public housing or the fostering or tolerance of informal tenures. The type of market can be strongly influenced by government action, and what may seem to be 'free' markets are often associated with strong state action, as the state makes the market in fundamental ways. A basic example is the need for markets to have a clear and binding system of contracts to enable exchange to take place. Markets need an institutionalised structure to enable them to work, and this is usually created and sustained by government action through legal mechanisms and the creation and regulation of institutions and their relationships. Some of this market-making may relate to markets in general, and some will relate specifically to the housing market, for example, the structuring of tenure forms that may vary over time and between countries and are set by government legal action. A further example is the legal structure of lending for owner-occupation and regulation of the lending criteria that lenders use. Actions of exchange professionals such as estate agents or real estate agents are also an important element of a market, and these are usually shaped by the type of market that the state creates. The structuring of the market by government can influence the actions of organisations within it and the market outcomes. Agencies in the housing market may be profit-seeking capitalist enterprises controlled by shareholders expecting high financial returns or, alternatively, collective, charitable organisations that are non-profit-making and controlled by their members who may seek a mutual or societal good in terms of housing output.

As Madden and Marcuse (2016, pp. 46–47) point out:

> The state cannot 'get out' of housing markets because the state is one of the agents that creates them. Government sets the rules of the game. It enforces the sanctity of contracts, establishes and defends regimes of property rights, and plays a central role in connecting the financial system to the bricks and mortar in which people dwell.

The state forms and regulates the financial system that can influence the way the market operates for individual households, but it also fashions the structure and behaviour of the market as a whole. For example, Ryan-Collins et al. (2017) argue that international regulatory moves since the 1970s have incentivised banks to favour property-related lending over other types of loans and so have contributed to the widespread increases in property and land prices brought about by this increased financialisation (see Chapter 8). The state may structure markets in different ways, and the consequence of the varying decisions taken by national governments is that housing markets differ in their structures and their consequent outcomes over time and between countries, as we shall see later in the chapter. Therefore, a government with a neoliberal ideology and discourse is likely to structure a market in a different way to a government with a social-democratic approach.

The construction of a housing market also has to take into account the nature of housing as a commodity. In other words, markets for different commodities may have to be constructed in different ways for effective functioning. The attitude towards these characteristics is shaped by the discourse used to understand them. There are different discourses used to understand the nature and functioning of the housing market. The traditional neo-classical economic approach, that shares similar assumptions to the neoliberal discourse, tends to downplay the differences of housing as a commodity and to see markets as universal phenomena. It is argued here that the nature of housing as a commodity means that markets will not behave in the manner that can be expected from the traditional neo-classical economic analysis and that this lies at the heart of the inherent contradictions and problematic functioning of the neoliberal housing regime. It is perhaps illustrative that many of the advocates of the adoption of behavioural economics, which attempts to move away from some of the more unsustainable tenets of the neo-classical approach, have used housing to illustrate their propositions.

Therefore, the chapter starts with an examination of the nature of housing as a commodity and the impact this has on the functioning of housing markets with a view to illustrating differences in the conceptions of the market and the consequent rationale for government actions in forming the market and of intervention within it. At the heart of this discussion is the importance of land in the housing market and the features that differentiate land from other forms of capital, the impact of which is to make neo-classical treatments of land and housing markets inappropriate and misleading. But even within the neo-classical paradigm, it is accepted that some markets may have imperfections, as well as externalities, that mean that individual costs and benefits may not align with social or environmental costs and benefits for the society as a whole. An example is the cost to society of segregation that may be created through the individual decisions of many people in their own interests, but may have the impact of undermining social cohesion or wider goals such as sustainability.

The chapter examines the implications of these different views of the housing market for some areas where government has intervened to 'make the market' such as the creation of tenure forms, the construction of housing transactions and the creation and regulation of a system of housing finance and mortgage lending. The implications of the different views are drawn out, and the increasing importance of the neoliberal discourse is shown.

The chapter continues with a review of the policy mechanisms that governments have to make and regulate the market and a discussion of the ways in which different housing markets can be evaluated, following the discussion of evaluation methods in Chapter 2. Some initial conclusions are made on the efficacy of the neoliberal housing regime that has been adopted in many countries.

The nature of housing

Housing is one of the most important uses of land, and land (understood as locational space and its use over time) is a major constituent of a house. Therefore, it is vital to understand the nature of land and the land market if we are to understand the nature of housing and the housing market. Ryan-Collins et al. (2017) note that land was treated as one of the three factors of production in classical political economy as in the work of Ricardo and John Stuart-Mill, but that, following the work of John Bates Clark, land became subsumed under the heading of capital in the neo-classical economic tradition. Ryan-Collins et al. (2017) argue that this is problematic and has been at the heart of the misunderstanding of the nature of land and housing markets inherent in this approach. Land is not like other forms of capital in that it is immobile; its supply is highly inelastic if not fixed (it is not possible to supply more, aside from some land reclamation schemes); it is eternal and does not wear out like other forms of capital; it is not homogeneous as every plot has its own unique location; and it is essential for all forms of economic activity. The permanence and inherent scarcity of land make it a good asset for the storing of value as it will appreciate rather than depreciate over time. The nature of land means that increases in demand cannot be reflected in increases in supply as in any other form of capital. Land can be moved between different uses but cannot be increased overall, and so prices will increase. Therefore, land is a major source of what economists have termed 'monopoly rent', that is the making of profit that does not depend on effort. The importance of this in economies today has been recognised in the concern with the rise in 'rentiership' outlined in Chapter 1, in which earnings from the increase in value of land and housing is a major constituent. Increases in the price of land are a major constituent of increases in house prices. A recent study of fourteen advanced economies found that 81 per cent of house price increases between 1950 and 2012 can be explained by rising land prices, with the remainder attributable to increases in construction costs (Knoll et al., 2014, p. 31). In the UK, the figure was 74 per cent. This feature explains why residential investment has taken an increasing proportion of asset wealth (Piketty, 2014).

Housing is a complex and unique commodity that has many different elements and so is difficult to encapsulate easily. Importantly, housing is both a consumer product that generates use value as well as a capital asset. Unlike many commodities bought in a shop, each house is unique in that it has a mix of attributes that may be similar but is never the same as others. For example, as in the analysis of land, each house has a unique location, and as we shall see in Chapter 6, location is an important element of the affordances of a house. But there are many other attributes of a house, as houses can also be homes, and as we shall see in Chapter 5, this means that many different aspects of identity and meaning get attached to the more functional aspects of a house such as the ability to access labour markets and public and private facilities such as shops, leisure amenities, health care and schools. A house is also an important positional good in many societies in that it signals status as well as access to resources. Therefore, housing can be important in issues of social cohesion.

A house is the major location for family life and the place where family members spend the majority of their time. It is a basic foundation for life, and there is extensive evidence of the importance of the home environment in influencing the educational achievement and health of family members. In addition, a house is a major family expenditure whether it is bought or rented and so has a large influence on living standards. Houses can also increase in value, and they represent a high proportion of household wealth in many countries.

Therefore, housing may be an important influence on the distribution of both income and wealth in a society, as housing is both an investment and a consumption good. Governments that have an interest in intervention in the distribution of income and wealth are likely to look towards housing as an important mechanism through which to achieve societal goals. To sum up, housing is a good that is:

- Unique (see Chapters 5 and 6)
- Expensive (see Chapter 8)
- An enabler for access to other resources (e.g. schools, health care etc.) (see Chapter 6)
- A positional good that signals social status (see Chapter 5)
- Where the form of consumption varies (e.g. between tenures) (see this chapter)
- Where consumption of utility and meaning is involved (affordances, see Chapter 5)
- Where individual costs and benefits may not align with social ones (see Chapter 11)
- Where outcomes matter to individuals (e.g. minimum physical standards, see Chapter 5)
- And to society (externalities, e.g. segregation, disease or social cohesion) (see Chapter 6, for example)
- Important in the distribution of income and wealth (see Chapter 8)

Housing as a market commodity

The unique qualities of housing mean that it is unlike most consumer goods or forms of capital. Housing is expensive and long lasting, and so households only transact in the market a small number of times, especially if they are owner-occupiers. The fact that the institutional structures of tenures determines the rights and obligations of residents means that the market is divided into, for example, rental and ownership categories that are separate in their everyday functioning, though strongly linked because prices in one will influence the prices in the other and these relative prices will influence the size of the respective parts.

The difficulties involved in a housing market are summed up in the two contributions considered below. First, Mowen (1987) identifies five factors that make up the traditional neo-classical approach to consumer behaviour, which are as follows:

- Rational behaviour
- Well-defined preferences and a knowledge of the satisfaction that will be gained from consumption of the product
- Perfect information about the product and market functioning
- Decisions subject to budget constraints
- Insatiable desires (i.e. more of a product will always increase satisfaction)

The important point here is that very few of these factors hold true for housing. For example, it is clear that much household behaviour in purchasing housing is emotional as well as rational. People make calculations about their housing choices, but also have emotional feelings about particular houses that may or may not feel like home. Preferences may not be well defined because there are so many trade-offs involved between the many different attributes of a house. It may be very difficult to get a perfect fit between preferences and outcomes, and there may be uncertainties in knowing whether someone is going to be happy in a particular house. Also, it is unlikely that households will have a perfect view of the future movement of house prices that can of course rise and fall over time and vary greatly between locations. It may be very difficult to tell whether a particular neighbourhood will become gentrified with the resultant price increases or deteriorate with price reductions. It is also possible that more housing may not increase consumer satisfaction because of its multi-faceted nature. What constitutes 'more' may vary between individuals. Even more in terms of physical space may bring problems of maintenance or cleaning. People may feel isolated or unhomely in a large space. Nevertheless, housing is also a positional good (Hirsch, 1977) that reflects status, and so consumption is likely to increase with incomes. It has been estimated in the UK that a 10 per cent increase in personal income leads to people spending 20 per cent more on housing space (Cheshire and Sheppard, 1998).

In the second contribution considered here, Maclennan (1982) identifies seven distinctive features of the housing market (1982, pp. 60–62) which render the use of standard consumer theory problematic:

- Individuals transact in the housing market infrequently, which means that consumers possess imperfect information regarding the state of the market.
- In the period between an individual's transactions, the market will have changed and evolved; therefore, any information that the individual possesses may be obsolete.
- Because it is costly to recontract in the housing market, imperfect information is likely to lead the consumer to engage in a costly search process.
- The fact that housing is a complex commodity exacerbates consumer problems in evaluating possible purchases.
- Evaluation is made more difficult by the spatially dispersed nature of vacancies.
- The process of house purchase entails engagement in some form of bidding.
- Because of the fixity of the second-hand housing stock, the relatively slow rate of turnover and relatively sluggish new supply, there is likely to be considerable disequilibrium in particular submarkets as a result of changes in demand.

These deviations from the neo-classical assumptions mean that some basic understandings about the behaviour of markets may not be appropriate in housing. For example, supply of housing may be inelastic in that it is slow to respond to changes in demand, a feature that has been shown in many different national housing markets (see Chapter 7). This means that markets may not 'clear' in the neo-classical sense, with markets able to stay in disequilibrium for long periods of time. Disequilibrium may mean long-term problems of housing shortage or unaffordability that mean that the market may not correct itself in the way that neo-classical theory would predict, thus opening the door or creating the perceived need for state intervention.

Maclennan's list also raises the issue of what economists call transaction costs. In other words, buying and selling houses can be a long and costly process with a number of agencies involved such as real estate agents, solicitors and banks. The efficiency of this process may therefore be important for households and for the functioning of the market itself. A long and expensive process may discourage transactions and reduce the responsiveness of supply to changes in demand. For many other commodities, transactions are regular and easily undertaken by consumers who may have lots of practice and become more efficient. This kind of learning is more difficult in the housing market where transactions are less frequent and transaction costs are high.

The transaction process varies between countries, partly because of different institutional structures brought about through path dependence, but also because of the impact of government intervention to make the market. As argued earlier, housing is a basic and costly commodity and the inherent risks in transactions for households are therefore high. This risk is compounded by the high transaction costs and by the asymmetries in information and power between consumers and other agents. For example, banks and real estate agents are involved in many transactions and can employ staff whose job is to build up expertise in being able to set house prices and to predict changes. Likewise, banks have staff with an expertise in predicting the future movement of both house prices and interest rates. Consumers are unlikely to be able to build up this knowledge, and the infrequent nature of their transactions means that any knowledge gained may be out-dated by the time they next transact.

There are different views of the nature of housing as a commodity and of the workings of the housing market. There is a neoliberal discourse, supported by reference to simplistic interpretations of neo-classical economics, that sees the market in similar terms as other markets and holds out the belief in a 'free' market that reaches equilibrium between supply and demand, based on a view that housing is not too different from other commodities. We have argued here for an alternative view of housing and housing markets based, to, some extent, on behavioural economics and on political economy analyses of the role of land in housing markets and in the economy more generally. This latter view is commensurate with analyses of 'rentier capitalism' outlined in Chapter 1 (and expanded on throughout the book). These two conceptions will lead to different conclusions made by governments on the appropriate shape and functioning of the housing market which will be elucidated in the rest of the chapter. The overall argument is that there have been strong moves in many countries towards the neoliberal conception, which has led governments to shape the market according to that ideology and discourse.

Constructing tenure

The chapter has shown that governments are deeply involved in the structuring of the housing market in many different ways. Perhaps the most basic arrangement is the legal structure of tenures. Tenures lay out the respective rights and duties of people and agencies that have an interest in a house. In the owner-occupation of a stand-alone house, these may be simple and uncontroversial as the owner is the occupier and perhaps the only other agency that has a right is the lender and then only in the circumstances of a default of payment. But even here, the state often has rights laid down in law that may set up a process by which property can be compulsorily purchased and the state may be able to enforce the undertaking of repairs if the condition of the building represents a danger or nuisance to others. In addition, tenure forms may have evolved that enable multiple interests in houses. For example, there has been recent discussion in the UK about the use of leasehold ownership of newly built houses. In these circumstances consumers purchase a long lease, and the freeholder retains an ongoing interest that is reflected in the annual payment of a 'ground rent'. The freehold interest can be sold on by the developer and in many actual examples was purchased by a company that sought to maximise its financial returns by massively increasing ground rents, resulting in leaseholders being subject to large and unexpected obligations that they had no control over. In this case, government intervened by outlawing the creation of leasehold in newly built houses.

The pattern of duties and rights is complex in multi-dwelling buildings where there are common elements of a building that are not the responsibility of the individual owner to

Figure 4.1 Multi-ownership of high-rise buildings

maintain and could cause problems if responsibility is not assigned. This was a problem in many East European countries with the privatisation of previously state-owned apartments in multi-dwelling blocks without a legal structure for the ownership of the communal parts, including the building structure. The result was a lack of responsibility for repair and maintenance and examples of individual action to solve collective problems. A striking photo shows the outside of a few apartments being clad to improve insulation whilst the majority are left in their original state (see Figure 4.1).

Many different forms of tenure have been devised to deal with the collective repair issues in multi-ownership blocks such as the US concept of a condominium or the Scottish Community-Based Housing Association. These two examples show the wide range of possible solutions that vary considerably on a continuum between individual and collective responsibility. In the Scottish example, responsibility for repair and maintenance of the common parts was vested in non-profit-making, collectively controlled organisations run by local residents. In contrast, condominium ownership in the USA was promoted strongly by housebuilding interests that saw the opportunity to access new markets and, originally at least, served their interests rather than that of consumers. See Case Study 4.1.

Case Study 4.1   Condominiums in the USA

The condominium can best be defined as a form of ownership of real property that combines an undivided interest, held in common with others, in land and certain portions of a building or buildings, on the one hand, with a separate and individual interest in a specific part of a building, on the other. The property owned in common is called the 'common elements', and that owned individually is called the 'unit'. The interest acquired by the condominium purchaser may be conveyed by deed or may be mortgaged as security for a loan, separate from the interests of the other co-owners. The unit is assessed separately, and the owner pays his own taxes.

Condominiums are a relatively new form of tenure in the USA that only got official authorisation when President Kennedy signed a housing act in 1961 which included in its Section 234 authorisation for the Federal Housing Agency (a state organisation) to insure first mortgages on one-family units in multi-family structures and an undivided interest in the common areas and facilities that serve the structure. But the concept needed to be sold to both consumers and, most importantly financial organisations that would provide the purchase loans. This selling was undertaken largely by builders and their representative organisations who could see a mechanism for reaching a market for low-income owners that had previously been difficult to access. The Federal Housing Agency itself, of course, had a definite interest in establishing clean titles and in maintaining value so that it could save on its insurance funds. Therefore, it specified restrictions on the projects that could be insured, and drafted a model statute to serve as an example for the state legislation needed for condominiums to proceed. The FHA then vigorously promoted the model statutes to ensure that the states would display a large degree of uniformity in terms of the description of the condominium, the procedures for establishing them, and the provisions for dissolving condominiums if necessary. Also included were guidelines for by-laws and other restrictions that would enhance rather than diminish the value of the property. The inclusion of recommendations for limits to liability, priorities of lien, and insurance coverage should be interpreted from the perspective of averting risk.

Given the importance in establishing condominiums by housebuilding and financing organisations, it is not surprising that the first generation of condominium statutes show a bias toward protecting the interests of the industry; they promote ease of procedures and protection of value. Consumer problems had not yet emerged, and thus the laws of most states recognised condominiums without attempting to regulate potential abuses. Many states left the matter of contractual obligations solely up to the buyer and the developer.

Many buyers' complaints were concerned with provisions written into the by-laws of the associations, and these could not be changed easily. Another problem was inadequate, incorrect, and incomprehensible disclosures on the part of the developers and their agents. And yet another was the failure of prospective purchasers to understand the costs and benefits of condominium ownership. Many problems were caused by the fact that the condominium brings together people with various ideas, lifestyles and interests in one homeowners' association. On top of this, various forms of abuse by developers were brought to the attention of public officials. These included business practices that are not illegal but are detrimental to consumer interests; breach of promise and of contractual obligations, even fraud. Frequently conflicts arose over the monthly assessments and over the developer's contributions to the association for the unsold units; this often led to unanticipated and substantial increases for the other owners.

Yet by far the greatest problem was the common practice of not including the recreational facilities among the common elements, but leasing these to the homeowners' association instead. This was done either directly, or through a special purpose company on a long-term contract (often 99 years); the fees increased rapidly due to contractually imposed yearly increases. Also, the costs of other services and provisions increased substantially, sometimes because of misrepresentation, sometimes because the developer stopped subsidising such provisions upon the completion of the sales. The disclosure of such problems helped bring about the second generation of condominium laws providing added protection for buyers in new projects and helping the condominiums regain some of their early appeal. But it took until 1980 before many of the old contracts could be changed.

The US condominium example shows the socially constructed nature of tenure relations and the impact of housebuilders, financiers and consumers on the tenure form and the role of government in responding (or not) to the competing interests and enabling the tenure form to arise and adapt.

Source: Based on van Weesep (1987).

In Sweden, the story of the co-operative tenure is an interesting case study in the way that tenure forms can be changed by government action. Originally formed by producer and consumer unions to own and manage the common parts of multi-dwelling buildings, they were originally collective and accountable in their structures and were set up to achieve collective benefit from any price rises. However, the legislation governing them was changed to allow dwellers to own their individual flat and to trade it on the market and to receive in full any increase in market value. In essence, a collective non-market good was turned into an individual, marketed commodity by state action. See Case Study 4.2.

Case Study 4.2 The marketisation of the tenant-owned apartment sector

The Swedish *bostadsrätt* sector is one of those idiosyncratic property forms that does not quite 'fit' conventional tenure categorisations – it is neither owner-occupation, nor rental. Instead, the occupier of an individual residential unit – normally, but not always, an apartment – is a member and shareholder of a co-operative (the *bostadsrättförening*) that itself owns the unit in question, together with one or more others (again, usually, but not always, in the form of a block or blocks of apartments). To acquire a share in the co-operative, a new resident pays a fee; in return, he or she gets the right to use a specific dwelling unit for an unspecified, unlimited period of time, and to transfer this right to a new resident/shareholder (Ruonavaara, 2005).

This sector – which I term the tenant-owned apartment sector – originated in the co-operative movement of the pre-war era, but it was not formally institutionalised until the period between the world wars. The two key pieces of legislation were the Tenant-Ownership Act of 1930 and the Tenant-Ownership Control Act of 1942. The latter was critically important insofar as it introduced regulation of transfer pricing: as a rule, 'the purchasing sum for right of occupancy was not to be higher than the tenant-owner's share of co-operative association's assets', with the latter 'calculated not on the basis of market price but the property's taxation value'. Moreover, local rent tribunals were to control transfers. The sector was thus located strictly *off* the market: it was a 'non-commodified form of housing' (Ruonavaara, 2005).

And it became, in the process, an increasingly integral component of the storied 'Swedish model', with the sector growing its share of Swedish dwellings from 4 per cent in 1945 to 9 per cent by 1960 (Figure 4.2).

As Ruonavaara relates, however, this sector also became, in the 1960s, the *first* component of the Swedish housing system to be deregulated and marketised and, thus, neoliberalised, in the manner envisioned by Wacquant. In 1968, the 1942 Tenant-Ownership Control Act was abolished, the principal upshot of which was a shift, from 1972 onwards, to market prices as the basis of transfer values (Ruonavaara, 2005)

In short, shareholders were henceforth allowed to sell their occupancy rights for the highest price that the market would deliver, and profit-oriented estate agents replaced public rent tribunals as the primary intermediaries in the exchange process.

Today, therefore, the tenant-owned apartment sector constitutes a subset of owner-occupancy in all but name – with the only material caveat relating, to restricted letting rights. By 1990, the sector's share of Swedish dwellings had increased to 15 per cent; by the end of 2010, it was estimated to have further expanded, to just over 22 per cent. Equally significantly, the deregulation of 1968 put in place the necessary conditions for two subsequent phases of runaway price escalation: first, from 1980 to 1990 (and lubricated by deregulation of Sweden's credit markets in the middle of the decade), during which period prices increased by 80 per cent in real terms (Turner, 1997a); and then, after interludes of falling prices (1990–1993) and relatively modest increases (1993–2000), from 2000 to 2010, during which period prices increased from a nationwide average of 390,000 Swedish crowns to an average of 1.3m – an increase, in real terms, of not far short of 200 per cent.

Source: Christophers (2013).

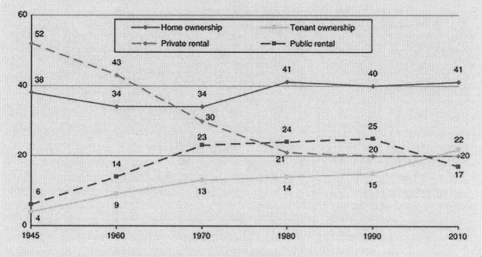

Figure 4.2 Swedish dwellings by tenure type, 1945–2010 (%)

Renting also has a legal structure that lays out the framework for determining the duties and responsibilities of landlords and tenants. These may vary considerably between countries and over time. In some countries, regulation is not well developed, and so the contract between the landlord and tenant assumes great importance. In others, there is a strong framework of national law that seeks to define responsibilities and protect the tenant against onerous terms and conditions as it is assumed that there is a power imbalance between the two parties. Common areas where governments may get involved include the length of tenancies, rent levels and the security of renters, including the circumstances in which the tenancy can be ended by both landlord and tenant. Also important, can be the fees paid to agents for finding lettings and discrimination against certain people such as benefit claimants or ethnic minorities. These are all issues that could impact substantially on the well-being of dwellers and on the nature of the tenure and its role. As a result, there are substantial differences between the tenure in different countries. In Germany, the private rented sector is closely regulated, comparatively large, houses dwellers at all stages of the life-cycle and is characterised by long-term tenancies. In the UK, the sector is loosely regulated, relatively small, largely houses dwellers in the early stages of the life-cycle and is characterised by short-term tenancies. The private rented sector tends to play a relatively minor role compared to owner-occupation in many regimes reflecting its weaker political position. In neoliberal regimes, the priority seems to be given to owner-occupation rather than private renting. In social-democratic regimes, priority is given to public rather than private sectors. It is only in conservative regimes that a regulated private rented sector seems to flourish. The differences in the construction of tenure rights can have important impacts on the lives of dwellers. If government does not intervene to balance the differences in power between landlords and tenants, then private renters can find themselves in insecure and unsatisfactory situations that do not seem to be righted by a competitive 'free' market.

In some countries such as India, the framework of rights and duties associated with tenure does not exist to the same degree, and where it does exist, many people may have to live outside it in squatter settlements with disputed or non-existent ownership rights. One of the main problems with slum upgrading programmes in these situations is the lack of legal title to the land that renders occupiers liable to lawful eviction at any time. In these circumstances, it is not surprising that many people find it difficult to take a long-term perspective and are reluctant to invest in their housing. Government action to upgrade properties or to stimulate investment is often predicated on action to establish legal occupancy rights for residents.

The social construction of the nature of tenure forms influences the experience that dwellers have in that tenure, but also the size of the tenure. Particularly when taxation and subsidy issues are included, government has a large influence over the popularity, economic cost and benefit of each tenure. Also important is the discourse about tenure. In neoliberal housing regimes, owner-occupation tends to be the dominant tenure and is 'normalised' through government discourse that stresses its strengths (Gurney, 1999). The discourse can be important in setting the balance between commodity and rights views of housing. A discourse that focuses on economic benefits and capital gain will tend to reinforce a commodity view of housing. There is a trend in many countries for tenures to be structured on a private, commodified basis as is shown by a number of case studies here. Neoliberal markets are deliberately structured to have the characteristics they exhibit, and this is not a natural or irreversible phenomenon.

Housing transactions

It was argued earlier that purchasing a house is a difficult transaction for consumers because of the need for substantial information for them to make a decision in awareness of the risks and obligations involved. Recognition of the imbalances in information has led some governments to regulate the provision of information given to potential purchasers. This may include the regulation of house descriptions, energy usage and so on. In Scotland, for example, sellers are required to provide, at their own expense, information to prospective purchasers on the structural integrity of the house and repairs needed as well as energy efficiency and the assessed value as a guide to the price.

There are different processes of contracting for house purchase. For example, in some countries, any prospective purchaser may make an offer that is accepted (or not) by the seller. In other countries, there may be a system of sealed bids in which the seller can name a date on which all bids will be examined and the choice made. Alternatively, there may be an open auction system in which bids are made and the highest wins. Each of these systems has different costs for the buyer and seller. In the sealed bids system, for example, prospective purchasers will have to engage in a search process to find and assess the property and to fund a building survey to assess the structural integrity of the house. This may have to be repeated many times if bids are rejected.

The search process undertaken will vary between individuals and the institutional structures and practices that exist in each country. In many countries, the process has been altered by the growth in internet use and social media. It is now commonplace to have video tours of properties posted online, and there has been a growth in the use of internet sites to advertise and sell properties both by the traditional agents themselves and sometimes as an alternative to the traditional system.

In private rental markets, there is also recognition by many governments of the imbalances in information and power between landlord and tenant in transactions in the sector. The regulation of deposits from new renters to landlords can be important, and there are examples of government action to stipulate the size of deposits and the conditions by which they can be withheld at the end of the tenancy. Some governments may provide funds to enable lower-income dwellers to afford a deposit and so enable access to the tenure. Another issue is discrimination by landlords against granting tenancies to certain people. Many governments have legislation to outlaw discrimination on the grounds of race or ethnicity, and concern is sometimes expressed if there is discrimination against people who receive state welfare benefits.

Also, agents in the letting process may charge for their services and government may think that fees are unreasonable for the work undertaken with the imbalance in power enabling some agents to charge a premium. Lack of competition among agents may have a similar impact.

Intervention in housing transactions may be justified on a number of grounds. In the neoliberal discourse, it may be justified in terms of making transactions easier and less costly and so reducing 'friction' in the market and increasing market sensitivity to demand. But intervention may also be justified in alternative discourses as a way to offset power imbalances and to improve the housing situation and the well-being of dwellers.

Regulation of lending for house purchase

The purchase of a dwelling is usually a relatively large expenditure for many households that is afforded through borrowing a proportion of the price from a lending institution. Therefore, the financing structure can have an important impact on the nature of the market and on the impact on individual households, and many governments have intervened through regulation to offset some of the more serious impacts on dwellers.

For example, regulation may be concerned to outlaw discrimination in lending practices either on the grounds of race or gender, or in particular areas, what was known in the UK as 'red-lining' where loans were not made available in certain areas considered to be 'high risk'. Regulation may also be concerned with factors concerned with individual loan transactions. For example, there can be substantial risk involved both on the side of the lender and the borrower. Regulation may be involved in the amount of money that can be borrowed as a proportion of the purchase price as well as general regulation of the amount lent by financial agencies.

Lenders may impose conditions on loans that may be considered onerous and borrowers forced to take out expensive and unnecessary insurance products. Some governments have recognised the uneven power relationships involved in house purchases and sought to protect buyers whether under general consumer regulation or more specific housing provisions. There have been many examples in the UK of the mis-selling of mortgage protection products by lenders, and they have been forced by government into recompensing borrowers.

Inability to repay a mortgage or loan for house purchase can result in the loss of a person's home. In some countries, there are regulations that encourage or require certain actions from financial agencies in order to attempt to prevent repossession of the dwelling. Also, some governments have been concerned to influence the type of loans available. In general, loans involve a mix of repayment of the capital and interest payments and can be for a substantial period of time (often 30 years or more) because of the substantial sums of money that can be involved compared to household annual income. Loans are generally one of three types, although there may be mixes. One is a fixed rate loan where the rate of interest is the same for all or part of the loan. The second is a variable rate that is usually set in relation to interest rates in the economy. The third is interest-only loans where the lender pays interest but does not have to repay the capital until the end of the loan. The distribution between these types varies between countries and government may intervene to encourage or discourage a particular type. For example, in Britain fixed rate mortgages have been encouraged and interest-only mortgages discouraged to help prevent problems of mortgage default. Variable rate loans can create problems for households when interest rates increase, thus increasing monthly repayments. Interest-only mortgages can result in the difficulty of repaying the capital at the end of the loan period if there is not an effective plan to ensure this capital is available.

In some countries such as Argentina, governments have been concerned about the availability of loans for house purchase. This may be because of the under-developed nature of the banking system or different priorities of banks. One solution has been to establish government agencies such as national housing banks to provide loans or to take government action to support existing institutions or to prioritise lending for house purchase. For example, in Argentina (see Case Study 4.3), the state has intervened heavily to increase the amount of lending for house purchase after economic crises reduced the supply of private finance and in a situation where inflation is very high, thus making long-term lending difficult for both lenders and borrowers. The government established a private trust underwritten by government to enable private funding to be provided for mortgages and established a system for index linking loans to inflation that had previously been illegal.

Case Study 4.3 Creating housing finance institutions in Argentina

Argentina has suffered from an enduring shortage of housing, and successive governments since the 1970s have reacted by providing funds to the devolved regional governments to provide loans to developers to provide new housing. The two major programmes, FONAVI (Fondo Nacional de la Vivienda, National Housing Fund) and Programas Federales (PF) began in 1972 and 2004, respectively, and both focus primarily on the supply side (i.e. loans for construction of new units, as well as infrastructure and home improvements, although they do also provide loans for the purchase of the constructed units). However, both have been criticised for being overly bureaucratic and for providing little housing for low-income groups.

Efforts to deal with the housing shortage have been hampered by two long-term problems in housing finance. The first is an enduringly high rate of inflation that has been running at about 40 per cent per annum. The second is a sequence of boom and bust economic cycles running throughout the 1970s and 1980s with two during the 1990s and an important one following the country's own default on international loans in 2001. The high rate of inflation has meant that renting in Argentina is considered to be a waste of money as rents increase annually with inflation, but ownership brings some stability to housing payments and ownership of an appreciating asset. However, the economic instability and endemic inflation have meant that the private sector has been reluctant to offer loans at the required level. The mortgage market in Argentina is still small when compared with other countries. According to country data from the Housing Finance Network and the Latin American Federation of Banks (FELABAN for its Spanish acronym), Argentina's mortgage market in 2012 was 1.5 per cent of GDP while in Panama and Chile the percentage is 25 and 18.7 per cent respectively, making these countries the largest mortgage markets as percentage of GDP in the region, surpassing also Brazil (6.8) and Mexico (9.1).

However, the extent of mortgage lending has been greater in previous years. For example, loans increased from practically being non-existent before to an equivalent of 5.7 per cent of GDP in 2001. Mortgage loans also became important for banks (27 per cent of their total loan portfolio) and for new homeowners, given that 25 per cent of new titles were financed by these loans. However, the loan repayment rate was low (between 35 and 26 per cent) and the macro-economic crisis of 2001–2002 resulted in a breakdown of the financial system and affected the availability of mortgages. But, despite the general economic improvement of Argentina during 2004–2008, the mortgage market never went back to its pre-crisis extent, representing less than 1.5 per cent of GDP on average.

Therefore, a new programme, Programa de Crédito del Bicentenario para la Vivienda Única Familiar (Pro.Cre.Ar), was created in 2012. In legal terms, Pro.Cre.Ar is a trust fund, an independent legal entity within the government, with the goal of facilitating access to a home for low- and middle-income families. Originally, the most important instruments of this fund were loans for the construction of residential schemes and also for the purchase of housing units, but Pro.Cre.Ar has been the focus of many changes intended to shift its focus to support the demand side by providing direct subsidies for individual mortgages, in contrast to the more supply-side orientation of FONAVI and PF.

> Pro.Cre.Ar is structured as a private trust and regulated accordingly. Therefore, it is outside the regulations applied to other operations of the state finances and investments, enjoying a higher degree of investment freedom. The life of the fund is 30 years since its inception in 2012 and is guaranteed by the national state.
>
> The fund comprises the following assets:
>
> 1. Land lots pooled by the Agencia de Administración de Bienes del Estado (AABE) – the agency in charge of managing the land and real assets of the state; as well as land contributions from provinces and municipalities across the country;
> 2. Funds made available by the national government; and
> 3. Debt issued by the trust
>
> The fund issues shares known as Certificados de Participación (CP –participation certificates), which are given to equity holders in exchange for land and funds (assets 1 and 2 mentioned above). The owners of these shares are the beneficiaries of the trust. The debt issued (3 above), known as Valores Representativos de Deuda (VRD – representative of debt value), are guaranteed by the state and can be traded in the open market. Currently, all the series have been purchased by the state's pension fund (Fondo de Garantía de Sustentabilidad), which is managed by the National Administration of Social Security (ANSES).
>
> During 2015 and 2016 the new government introduced a number of changes to Pro.Cre.Ar. A system of direct cash subsidies to the demand side was implemented in order to boost the mortgage market and to replace the old system of indirect supply-side subsidies. The demand-side subsidy is funded by the Federal Government and distributed by Pro.Cre.Ar. Additionally, loans indexed to inflation, which had previously been illegal, were allowed by government, and commercial banks now lend inflation-adjustable mortgage loans to households and at market interest rates.
>
> This case study illustrates the importance of government in making the housing market in the specific situation faced by individual countries. At the same time, it shows the way that financialisation is becoming and established trend even though in Argentina it is starting from a very low base.

In the predominantly neoliberal policy environment, lending for residential purchase has become an increasingly important element of overall financial markets. Ryan-Collins et al. (2017) have argued that the importance of land as a store of value, outlined earlier, has meant that banks have been increasingly attracted to lending that is secured on land and have become essentially 'real estate lenders'. The result has been an increased 'financialisation' of housing to varying degrees in many countries.

> Financialisation is a term used to describe the penetration and increasing influence of financial markets, motives, institutions and elites into new areas of the state, economy and society. Financialisation involves the transformation of work, services, land or other forms of exchange into financial instruments, for example, debt instruments like bonds, that can be traded on financial markets.
>
> (Ryan-Collins et al., 2017, p. 120)

Traditional lenders for house purchase, such as British Building Societies, depended on savings to generate the asset base for lending and so were restrained in what they could lend by the amount of savings and by income growth to enable this personal saving. However, the growing liberalisation of financial markets pursued by many governments and international institutions has led to the situation where mortgage lenders can borrow themselves in order to provide loans and so are not restricted by growth in incomes. In effect, they create money in a 'house price/credit feedback cycle' in which new loans on property lead to increasing house prices which, in turn, provide security for future loans. This increased lending leads to increasing demand and prices in the housing market.

Ryan-Collins et al. (2017) argue that banks create credit and money against existing land, as with the majority of home mortgages, and that this can inflate house and land prices. Rising prices mean households and firms must borrow more to become home- or landowners. Therefore, the supply of bank credit can be seen to create its own increased demand for even more credit, assuming a fixed supply of land. The demand for land can then stretch well beyond people's incomes or firms' profits – and the growth rate of the economy – in particular if people expect house prices to continue to rise faster than incomes.

Also, in order to hedge against risk of default, lenders may wrap up loans into securities that may be sold on to other investors. Securitisation enables mortgage issuers to offer a wide range of mortgage products, to offer mortgages at much lower rates of interest and offer them at higher loan-to-value (LTV) ratios. This in turn enables larger numbers of people to access home ownership at higher price-to-income and mortgage debt-to-income ratios (Ryan-Collins et al., 2017). However, this secondary mortgage market was the cause of the financial crash of 2007 when the risk associated with these securities was systematically undervalued, and so institutions came under pressure when the level of default increased along with an economic recession. See Case Study 4.4.

Case Study 4.4   The Global Financial Crisis

The roots of the Global Financial Crisis can be found in the rise of the secondary mortgage market in the USA and in other countries (such as the UK) in the 1990s which was created by government through institutions such as Fannie Mae and Freddie Mac. Fannie Mae was a public organisation at first, but was later privatised, and Freddie Mac was established as a private organisation, although both had some degree of government underwriting of risk. At first, secondary market institutions purchased the mortgages originated by mortgage banks and held them in their own portfolios. However, starting in the 1960s but growing quickly in the 1980s, the secondary institutions started to aggregate these mortgages into various types of financial securities such as bonds. In the 1990s, a growing number of investment firms joined the secondary mortgage market, but unlike Fannie Mae and Freddie Mac, they did not insist on as tight underwriting standards (for so called 'prime' loans) and so specialised in higher-risk or sub-prime loans. The rise of these securities linked the housing market into the general finance system as many investors could purchase mortgage bonds as they would any other type of bond. By the early 2000s many bonds were in the hands of foreign investors. By the mid-2000s bonds became increasingly sophisticated and the link to the originating high-risk mortgages lengthier. At the same time, the credit rating of these bonds became problematic and considerably understated the risk.

The growing use of bonds coincided with a rise in 'risk-based' pricing of mortgages. In other words, rather than turning down applications for riskier mortgages, lenders just increased the interest payments. These became known as sub-prime mortgages and were mainly offered by non-depository institutions, that is those who raised their income from loans from other institutions rather than from deposits and savings. In the view of Schwartz (2015, p. 87), 'a significant proportion (of these loans) was predatory – conducted with little or no regard to the borrower's ability to afford the loan, and often involved deceit and other forms of fraudulent behavior', This made them vulnerable when the supply of loans dried up when the financial system came under pressure and mortgage foreclosures increased.

The rise of sub-prime loans was enabled by government through legislation that deregulated the market in the 1980s and the consequent refusal to regulate them during the 1990 and early 2000s. Also, mortgage lenders were motivated by increasing or stabilising their share of lending at a time when there was fierce competition.

Schwartz (2015, p. 95) likens the housing finance system in the USA in the mid-2000s as a sandcastle that was very intricate and elaborate in its design, but devoid of structural integrity. He argues that the system was predicated on the patently false assumption that house prices would always increase and fell apart as soon as house prices started to fall in 2007. Failure of mortgage repayments, that in turn led to rapid increases in foreclosures led to investors in sub-prime mortgage backed securities to leave the market. As a result, secondary institutions cut back their acquisition of sub-prime loans that left the lenders in the lurch with many becoming bankrupt or leaving the business. The repercussions resonated through the global financial system as many of the largest institutions experienced a liquidity crisis and some collapsed, most famously Lehmann Brothers in 2008.

Although federal and state legislation existed to prohibit predatory lending, in practice flaws in the system and the lack of enforcement meant that they were not effective. It was only with the Dodd-Frank Wall Street Reform and Consumer Protection Act of 2010 that a number of measures were enacted intended to prevent a mortgage crisis happening in the future. As for the secondary mortgage market, Schwartz concludes that government has been slow to agree and implement a new system to replace the old.

Note: For a fuller account, see Schwartz (2015).

There is variation in the extent and depth of financialisation in mortgage markets in different countries. There is evidence that the more liberalised the financial regime, the more volatile is the housing market and the stronger the relationship is between house prices and the wider economy (Ryan-Collins et al., 2017). In other words, the housing market tends to destabilise the overall economy.

It is important to reflect that financialisation of housing is not a natural force, but has been created by the type of and lack of financial regulation of credit in general and house loans in particular. Ryan-Collins et al. (2017) show how the regulatory situation in the UK changed during the 1980s and after to create the present liberalised situation with its resultant problems of affordability and volatility that reflect the underlying nature of land and the land market outlined earlier. They also outline regulatory and taxation policy tools that could be used to change this situation which will be considered more fully in Chapter 8. The example of British Building Societies shows the importance of these changes and the impact of government action in enabling the societies to change from mutual non-profit organisations to profit-making private organisations integrated into the wider financial system. A change that has resulted in alterations to their modes of operation that have profound implications for dwellers who seek to enter owner-occupation as well as the nature and characteristics of the tenure itself (see Case Study 4.5).

Case Study 4.5   British Building Societies

The institutional structure of the British owner-occupied sector has historically been based around the building societies (for a history of the building society experience see Boddy, 1980). They started in the early nineteenth century as self-help societies primarily formed by skilled working-class people in order to organise the building and financing of owner-occupied housing. These societies were temporary in that they were disbanded once all the members were housed. They were also based on a direct link between saving and being housed. Both of these features disappeared during the nineteenth century as the sector grew and became subject to government regulation through a series of Acts of parliament. By the late 1980s, building societies had become major financial institutions with a clear role and a close relationship with government.

The state has played a key role in constructing the finance system. This was evident in the early years of the building societies and has been continued up to the present day. It is illuminating to compare the private housing finance sector in the mid-1980s with the sector in the 1990s. This shows how the sector has been radically refashioned largely through government action.

In the mid-1980s there was a very distinctive and internationally unusual institutional pattern that involved the building societies and the government. There was a large number of building societies, although a few of the largest dominated the market. The organisations had a mutual ethos, being owned and controlled by their members who were their depositors or savers and borrowers. However, accountability to the membership was not strongly developed. Essentially the societies were run by their senior management with little control from the membership. The societies were conservative in their operations, generally sticking to tried and tested practices with little innovation. They were risk-averse with stability a major organisational objective. There was a close relationship with government, which regulated their activities through a legislative framework that had as its cornerstone the protection of borrowers and savers. Therefore, it laid down rules on the financial structure of societies and limited their role to the housing sector. Within this sector the societies were given taxation advantages over other financial institutions, making it difficult for their dominant position to be challenged by, for example, the high-street banks. Competition between societies was gentle and limited in scope. An effective cartel of the large societies set interest rates. Therefore, price competition did not exist and rivalry for customers was never of primary importance and was restricted to general image promotion and competition for prime high-street sites. Coupled with managerial control, this resulted in luxury branches with large marble halls!

Government used the societies as one of the major instruments of macroeconomic policy. Through the Bank of England, it controlled the amount of money the societies could lend by stipulating the amount they had to deposit with the central bank. This was one part of a wide range of controls on credit and on financial transactions such as currency exchange, which supplemented interest-rate changes. When it wished to control credit to reduce economic activity and restrain inflation, the Bank of England would restrict the amount of credit the societies could offer and, conversely, would increase credit availability when it wanted to increase economic activity. This resulted in a relatively stable housing sector with credit being expanded when house prices were falling and restricted when prices were increasing. In this way fluctuations in house prices were evened out and this helped to stabilise fluctuations in consumer spending and general economic activity.

The relationship with government was a cosy, corporatist arrangement that suited both parties. The societies received tax advantages and a protected environment in which competition was limited and entry to the sector difficult. The government turned a blind eye to interest rate collusion and in return had an effective tool in the management of the national economy. Consumers had the advantage of a stable housing market, but the risk-averse behaviour of societies, together with the credit controls, meant that the availability of mortgages was limited. There were queues for mortgage lending and the building societies used rationing criteria based on a household creditworthiness and savings record as well as the location of houses. Housing research studies in the 1960s and 1970s drew attention to the discriminatory behaviour of societies in denying loans to people in areas that were perceived as at risk of house price decline or who, because of their ethnic origin or economic position, were considered to be at risk of not meeting repayment obligations. The red lining of some, usually inner-city, locations was held to be a major factor in the decline of some parts of cities (Boddy, 1980).

In the 1980s a number of factors influenced the government to change the sector. The increasing globalisation of financial capital resulted in pressure from the financial community to lift controls on the sector to allow it to compete with other world financial centres. Banks became increasingly critical of what they saw as the favoured position of building societies and demanded access to the housing sector. This coincided with the election of Conservative governments from 1979 onwards that discarded the traditional Keynesian techniques of economic management, which included the housing stabilisation arrangements, and instead put their faith almost exclusively in control of the money supply to manage the national economy. These Conservative governments were also committed to owner-occupation and so perceived the rationing behaviour of societies as restricting the growth of that sector.

> Therefore, a new institutional system was constructed during the 1980s by the government. Barriers to entry to the housing finance sector were reduced and, in return, restrictions on building society activities were relaxed and societies given the option of turning themselves into public limited companies along the lines of the existing high-street banks. The system of colluding on the setting of interest rates was outlawed and competition encouraged. The system of credit controls operated by the Bank of England was dismantled as part of a general deregulation of the financial sector. Government relied on the policy instrument of interest rate changes to regulate both the housing sector and the wider economy.
>
> One result of these changes was the entry of new players into the housing finance sector. This mainly consisted of the high-street banks, but also included, during the 1980s, direct lending institutions that did not rely on consumer deposits to provide the funds for mortgage lending. Instead, funds were borrowed directly from the money markets. Competition increased and interest rate variation between lenders appeared. Many societies took advantage of the opportunity to shed their mutual status and become private banks. This change was achieved by buying off the mutual owners through the provision of cash payments for their loss of ownership (see Stephens, 2001).
>
> This example serves to show that there is nothing inevitable about the structure or practices of housing finance institutions. It is sometimes assumed that markets in general have a common and predetermined structure that is implicit in their status as markets. However, markets are created and sustained through human agency in different circumstances. A key player in this construction is the state, which creates the legal and institutional framework within which markets operate.
>
> Source: Clapham (2005).

Policy tools

This chapter has demonstrated that housing markets are made by governments that set the rules, formulate the nature of the housing commodity through the construction of tenure forms and shape and influence the main institutions and actors. The examples shown in this chapter illustrate some of the policy mechanisms that can be used to control housing markets. The example of US condominiums shows the importance of government action in legislating for particular innovations in the housing regime. This function is paired with the creation and design of organisations and institutions such as the Federal Housing Agency to guide implementation of the innovation by working through detailed implementation and guiding other agencies, in this case through the provision of model rules for them to implement.

Also, the emphasis in this chapter has been on the regulation activities of government in enabling the market to function in a particular way and in protecting consumers and others from market failure. The example of the regulation of finance for house purchase shows the difficult task facing government in a financialised regime. Previously used policy tools, such as credit controls, seem very difficult to use effectively in a housing finance system integrated into global finance markets. Also, the replacement by government in the UK of non-profit-making building societies with profit-oriented banks has meant that the regime is made up of market-oriented organisations that are more difficult to control and regulate. The options available to governments seem to be narrowed considerably in a neoliberal, financialised regime.

However, effective policy tools do exist, such as taxation arrangements to restrain price increases as we shall see in Chapter 8. The implementation of these tools is dependent on the outcome of the language games introduced in Chapter 2. In making the market in a particular form, governments are also taking part in the language games that will frame what is considered normal and expectations of what is possible. Sandel (2013) has noted the impact of the market on many aspects of people's lives and how this influences values and shows itself in personal lifestyles. The construction of a market that delivers large capital gains to many individuals fosters the idea that housing is a commodity with a value, rather than a necessity of life that is prized for its use value. The examples in the chapter show how collective, mutual and communal organisations and institutions have been altered in neoliberal regimes to fit the market commodity ethos. In addition, this approach creates substantial coalitions of interest around an owner-occupied market with substantial price increases that benefits existing owners as well as exchange professionals and producer interests, including landowners. The losers from such a regime are those looking to enter the sector and renters, but in neoliberal regimes, these interests seem not to be able to change the dominant market discourse.

Evaluating markets

Governments shape the housing market and the uniqueness of housing as a commodity means that the functioning of the housing market does not follow the perfectly functioning equilibrium models of neo-classical economics. There are longstanding market imperfections, externalities and high transaction costs that give reason for state intervention to rectify imbalances in information and power and to improve market functioning. So how do we judge the efficacy of market functioning, and how do we evaluate the interventions of government?

As outlined in Chapter 2, there are two major criteria for evaluation, either on the basis of the procedural forms of market processes or their outcomes. The first would involve assessment of the processes involved in the functioning of the market by a number of yardsticks. It was argued in Chapter 2 that there is some disagreement among neoliberal writers about the importance of competition as a market principle. In some readings of the neoliberal paradigm, markets are held to function most effectively where there is competition between producers and between consumers that is said to lead to maximum utility as well as the moral virtue of liberty. Therefore, government pursuit of competitive markets has become a major activity of some neoliberal states as shown in anti-trust measures in many fields, although the dispute within the neoliberal paradigm has meant that implementation has not always been effectively sought or achieved by neoliberal governments. In the housing market, competition seems to be equated with a free market in housing supply, and we will return to this in Chapter 7. In addition, competition has been interpreted as reducing government restrictions on the activities of market agents such as financial organisations as shown by the action on building societies in the UK in which their freedom to change their organisational form and to remove their perceived competitive advantage against banks was justified on the basis of the competition discourse. Another appropriate market criterion may be the size of transaction costs. The lower the transaction costs to both sellers and buyers, the more transactions there are likely to be in the market thus aiding the sensitivity to consumer demand.

The second way of judging housing markets would be on the outcomes for dwellers. For example, government interventions designed to remedy information and power imbalances can be evaluated on distributional grounds as they are sometimes justified on the basis of preventing discrimination or creating a more equal playing field for different groups of people. We will examine these kinds of interventions in more detail later in the book. However, even at this stage we can register some warnings about the outcomes of a neoliberal regime with its market domination and financialisation and the increased risk of volatility and house price increases that can lead to problems of affordability which will be discussed more fully in Chapter 8. House prices in the neoliberal UK have been increasing at an average of 3.6 per cent from 1970 to 2015 compared to 1.5 per cent in social-democratic Sweden and −0.3 per cent in conservative Germany (Meen, 2016).

Stephens (2011, p. 18) defines volatility as 'rapid fluctuations in house prices'. While this affects many countries, the UK has one of the most persistently volatile housing markets, experiencing four major boom and bust cycles since the 1970s. In addition, the two most recent property crashes have involved falls in the nominal (cash) as well as the real (inflation-adjusted) value of houses. Because mortgage debt is fixed in nominal terms, the fall of house prices on this measure is particularly damaging as it can cause homeowners to fall into negative equity. Case Study 4.5 shows how the increased volatility of the UK housing market has been associated with deregulation of housing finance and increased financialisation, involving greater integration of housing finance into the wider financial markets. At the same time, traditional housing policy mechanisms such as credit controls have become difficult to implement in this new regime.

A volatile market has a number of profound impacts on the outcomes of a housing market and on dweller well-being.

> For example, large price increases distort a household's specific housing choice, decrease housing affordability, affect house building decisions and intergenerational equity as well as lead to problems with low-income households being unable to access suitable affordable housing. On the other hand, a sharp fall in housing prices would cause issues such as negative equity, mortgage defaults and associated levels of instability for households seeking to attain homeownership.
>
> (Lee and Reed, 2014, p. 1074)

However, the extent of volatility is not the same in other neoliberal housing regimes. For example, although the level of volatility is considered to be an increasing problem in Australia, it is much less of a problem in the USA (Oxley and Marietta, 2010). Nevertheless, the US housing regime was responsible for the Global Financial Crisis in 2008 and its impact throughout the world. (See Case Study 4.4.) Social-democratic regimes and those with smaller owner-occupied sectors are less likely to endure high housing market volatility. The reasons for high volatility are probably complex and embedded in the market and taxation structures of each national housing regime. However, there is little doubt that neoliberal regimes are more susceptible to the problems that can arise from this.

Governments may defend their policies designed to increase the size of the private sector, and particularly the support of owner-occupation, as increasing the housing well-being of dwellers in owner-occupation. But, Matznetter and Mundt (2012, p. 289) point out what they label 'an interesting conundrum':

> While homeowners in most countries usually show a *higher* level of housing satisfaction than renters (Elsinga and Hoekstra, 2005), countries with a high rate of home ownership usually show a *lower* general level of housing satisfaction than countries with large rental segments (Czasny et al., 2008).

In their study of EU countries, Czasny et al. (2008) divided countries by welfare regimes and ownership shares. Countries with low ownership shares generally corresponded to the conservative and social-democratic welfare regimes. These countries performed better in terms of household satisfaction with dwelling and living area than high-ownership countries. Of course, many factors could be influencing these findings, and it is not known how people make judgements of satisfaction and who or what they compare their situation with. Also, there are many other factors that could be used in judging outcome such as the proportion of people who are homeless. Nevertheless, the important issue here is that the outcome of housing policies and housing regimes is an important element that should be taken into account in looking at different countries.

Clapham et al. (2017) shed some light on the processes involved in differences in well-being by showing that households moving into owner-occupation report an increase in their well-being. However, this is compensated, at least in part, by the reduced well-being of others. As Clapham et al. (2017, p. 14) argue

> The crucial point is that if the effect of home-ownership on subjective well-being is mediated by social status, then increasing rates of home-ownership may increase status and subjective well-being of first-time buyers, but this will be at the expense of existing home-owners and those renters left behind, both of whom will suffer from reduced social status.

Existing owner-occupiers will find their elite status diluted and renters will experience a decrease in their numbers, thus reinforcing their marginal status.

The increase in owner-occupation is usually among those who are marginal owners, that is households that can find it difficult to sustain the financial payments to sustain ownership. Haffner et al. (2017) argue that, if such marginal owners are forced to leave the tenure, their well-being reduces substantially. Interestingly, they note that there is a rebound effect as households adapt to their new status, but that this effect is much greater in the UK than in Australia. They explain this by arguing that UK leavers are more likely to gain access to the public housing sector, whereas in Australia, leavers do not have such a 'soft landing' and so experience a more longstanding reduction in their levels of well-being.

The above discussion shows that the make-up of the housing regime may systematically impact on the well-being of dwellers and that, in making the market, governments are influencing the well-being of their populations. There is some, although limited, evidence that neoliberal regimes with their large owner-occupied sectors may result in lower overall well-being of their populations.

Conclusion

The chapter has shown the unique nature of housing as a commodity that makes it problematic for market exchange as it does not conform to the neoliberal and neo-classical understandings of how a market operates. Many of the arguments for government intervention in the housing market that we will examine in future chapters arise because of the characteristics of housing that mean that housing outcomes are not always acceptable to dwellers and governments.

Also, the chapter has shown that government intervention is needed to establish and maintain the housing market. The market depends on government to make and police the rules that enable exchange to take place. Therefore, government intervention in housing is extensive in most countries. The national markets of different countries may vary considerably, and some of this difference may be due to the path dependence of institutions, but some is due to the actions of government in shaping the market to conform to different ideologies and discourses.

Examples have been given in the chapter of ways that governments construct the market, through constituting tenure forms and ordering the exchange mechanisms by, for example, rectifying information and power imbalances in the markets for ownership and rental that some governments seek to assuage. This may be done through general consumer protection legislation or specific housing regulation. The usual aim is to reduce risk in the market for consumers and to aid specific groups to access the market. Governments, also, structure and regulate market actors, and the example of the regulation of financial institutions is given in the chapter. The discussion of financial regulation highlights the increasing financialisation of housing regimes in many countries, and this chapter shows the impact of this on the policy tools available to governments and on the volatility and price changes in the markets. These detrimental outcomes are more likely in neoliberal regimes, and we will return to this in future chapters.

As Haffner et al. (2017, p. 1094) conclude,

> If the test of a well-functioning housing system is the wellbeing of its occupants, the findings of this paper present a challenge for regimes anchored on owner-occupation. The edges of ownership are too broad, and the path to outright ownership too precarious, for home ownership to retain its reputation as a crucible of wellbeing. Institutional differences may inspire sustainable solutions within jurisdictions, but cross-national convergences dominate the findings, and they question the therapeutic qualities once ascribed to ownership-centred housing systems in the English-speaking world.

Chapter 5

House and home

Introduction

The aim of this chapter is to describe the constituents of a house and home and the focus and impact of government policy on this area. It is important for housing policy to understand the nature of housing, because it underpins the experience of dwellers and the impact of housing policy. It is interesting that, in the English language, it is common to talk about a home as being the place where people live, but the policy directed towards it is labelled housing policy not home policy. Some people have argued that this disjuncture is at the heart of many housing problems today as policy-makers focus on houses and not homes (King, 2009; Clapham, 2018). However, a housing policy focused on homes is likely to have a positive impact on dweller well-being.

The analysis here uses the concept of 'affordance' (introduced in Chapter 2) to provide a framework that enables analysis of the multi-dimensional concept of home that includes both the physical structure and the meaning. Affordances of an individual in their home environment are structured into residential social practices that are modes of behaviour within the home that include basic living activities as well as other elements of home life that have a spatial and time dimension. Residential social practices are undertaken to enhance an individual's subjective well-being, and this concept provides an important measure with which to evaluate the impact of government intervention in this area.

The key questions that are asked in the chapter are what is a 'good' home and how effective are government policies in achieving this. The concept of subjective well-being (introduced in Chapter 2) will be used to help answer these questions. The house and home are the foundation of all housing policies as they are the basis of the dweller experience. At the same time, a house or home is itself the direct focus of government policy. For example, many governments intervene in the market to attempt to improve the quality of housing. This may be through standards set for new-built housing, legal standards for the adequacy of existing housing because of concerns about the lack of facilities or overcrowding, or through the provision of aid to enable households to keep housing in good repair. However, government interventions may have consequences on the market, and one result of intervention in housing standards may be an increase in costs for developers, landlords, owners and residents.

The chapter begins with a review of the affordances of home and the implications this has for the definition of a 'good house'. This forms the basis for a discussion of the reasons for government intervention in this area and the mechanisms that are commonly used. The discussion draws out the influence of the housing regime type on both the incidence of poor-quality housing and the probability and extent of government intervention as well as its rationale. Finally, some conclusions are drawn on the impact of government intervention.

The affordances of home

The concept of affordance stems from the work of Gibson (1986) in perceptual psychology. The essence of the approach is that people see their environment holistically on the basis of what it affords them, that is what it enables them to do or feel. Gibson argues that people see the physical or material aspects of an object at the same time as its use and meaning. For example, in the UK, if someone sees a red metal object by the side of the road they perceive it as a post box that has the affordance of receiving letters to be sent. But use has both a practical and a meaning component, which is vitally important in assessing the impact of housing as both aspects are inherent in the home. A house provides shelter, but also many emotional and symbolic elements, as we shall see in the next section. The concept of affordance can be used to assess a house or home holistically and so enables us to assess its impact on the well-being of residents, which is at the heart of the assessment of housing policy.

The affordances of a dwelling for a resident may be formalised into a set of residential social practices. These are the activities that take place in the setting of the house and may be routinised in time as well as in particular interior spaces. Different physical environments will afford particular social practices, and these will vary between individuals according to attitudes and lifestyles. Therefore, the concept is subjective. Within a household the practices will be negotiated between members and will reflect power structures within household relationships, which will be influenced by attitudes towards gender and age-related roles. For example, whose role is it to clean the house or wash the dishes? Clearly the availability of technology and of paid help will influence the answers to these questions as well as general social discourses and norms. Another example is whether children should be seen and not heard in the Victorian fashion or at the centre of family life as is more usual in households today. Residential social practices will reflect these discourses and values.

Residential social practices cover all aspects of home life. From a policy perspective, some may be more important than others because they include the basic functions of life such as washing, eating, or sleeping. There are measurement tools such as Activities of Daily Living (ADL) scales designed to help practitioners such as occupational therapists to identify and rate the ability of an individual to achieve these activities in a particular environment (see Clapham, 2017).

What is a good house?

The physical structure of a house affords many uses and meanings. The design and layout of the house may impact on these affordances at many levels by influencing the way that people live and by structuring household activities. Of course, the opposite causality is also important, as households will alter and shape the house to fit the desired household practices. This two-way relationship is shown by the different forms that houses take in different cultures. For example, in his study of the Berber house, Bourdieu (1970) shows how it is designed to enable spaces to be defined for specific gender use, that is there are male and female spaces within the house that reflect cultural norms about gender divisions of labour and the mixing of genders. So, the desired internal layout of a house may vary between different countries and cultures reflecting different household and family structures and social norms and practices. There is a time element here also in that these social practices may change over time. For example, much has been written about the changing size and layout of kitchens and their place in the house. In the UK, they have moved from the back of the house in Victorian times where they were shut out of sight and away from public gaze and the preserve of servants in middle-class households, to the centre of the contemporary house and the site of many household activities such as eating and socialising as well as cooking and washing clothes and dishes. Contemporary houses may have many bathrooms that have become places of body care and worship, whereas in many countries, it was common for houses not to have bathrooms, as washing was undertaken in more generic areas of the house. Time is also important in that the domestic practices in a house may vary over time. A teenager's bedroom may be a semi-public space to socialise in during the day, but a private space for sleeping at night. Spaces in a house may be divided between front and back regions depending on whether they are open to the gaze of visitors or private. Some areas or rooms may afford public display of objects designed to show the status, identity or lifestyle of the resident, for example, through the display of books or porcelain or other display objects. Bourdieu (1984) has written about the status importance of taste and house furnishing and decoration is an important arena for taste to be displayed and social status gauged.

The form of houses has meaning for those who live in them, but also for others. Research in the USA shows that people make assumptions about who lives in a house from its appearance (Nasar, 1993). Houses that are the same except for the external detailing can give different messages to passers-by about the people who live there and their perceived affluence, friendliness and so on. Also, houses can have a wider symbolism reflecting discourses in the wider society. Case Study 5.1 on British public housing shows this clearly, as societal discourses become embedded in the physical form and meaning of the housing.

Case Study 5.1 The symbolic meaning of housing: Public housing in the UK

The contrast is often made between public housing built in the UK in the 1920s and 1930s. In the 1920s, new-built social housing was homes-fit-for-heroes returning from the First World War and was aimed at skilled workers and middle-class clerks and teachers. The houses were designed to look like rural cottages and set out in tree-lined avenues and cul-de-sacs with large elements of green space in a re-creation of the traditional English vision of a rural idyll. In contrast, in the 1930s, the policy emphasis switched to the replacement of slums and catering for the poorest sections of the population. House design reflected this with properties in blocks of apartments with little external detailing – rather barrack-like in appearance. Over time, the 1920s estates have retained their high status (although many have been sold under the right-to-buy) whereas the 1930s estates have remained unpopular and stigmatised. The external appearance of the houses reflected the aims of the builders as well as the perceived status and importance of the residents in a self-reinforcing cycle.

> Ravetz (2001) sees council housing in the UK as a cultural colonisation in which a vision was forged by one sector of society for application to another. Cole and Furbey (1994, p. 112) argue:
>
> The spatial segregation and architectural distinctiveness of many estates, most notably the post-war flatted developments, involved the powerful imposition of producer's meanings and symbolism on working class households. Such architecture announces to society the 'differentness' of the scheme's residents, underlining the marginal status of many inner-city households and prompting the question, in the context of concern about 'problem housing estates' as to why people should be expected to care for the symbols of their own social inferiority.

But the meaning of a house is not necessarily fixed forever. We shape our buildings, and afterwards our buildings shape us. A house can be seen as the object of human agency *and* as an agent of its own through its relative fixity. A house is a site for people and organisations to define themselves and pursue their goals, but also one where those meanings and purposes get structured and constrained. Gieryn (2002) identifies three important moments in the life of a building. The first is the design where the building takes on the meaning of the designers; the second is the shift of agency to the building that structures human relationships; the third is reconfiguration through narration and reinterpretation by users.

The physical structure of the house and the meaning it has for residents are inter-related. Some studies have focused on the everyday discourse of home where the aim has been to attempt to describe and analyse the meaning that the concept has for individuals, groups and societies. In some cases, this has taken the form of lists of meanings. For example, Somerville (1992) uses six categories of home: shelter, hearth (by which is meant feelings of physical warmth and cosiness), heart (loving and affectionate relationships), privacy (power to exclude others), abode (place to call home) and roots (individual's source of identity and meaningfulness). This list seems to be blind to cultural aspects of shared meanings, but other lists include cultural as well as experiential elements (see, for example, Després, 1991). Therefore, it is recognised that different aspects of home may be important in different societal contexts.

Homes have important emotional and psychosocial impacts. Clark and Kearns (2012) found a strong correlation between high mental well-being and four subjectively assessed elements of a person's house that relate strongly to the factors highlighted as being important to general well-being: their view of its external appearance and what other people thought of it (social status contributing to self-esteem), a feeling of being in control of the home (personal control), the home expressing one's personality and values (identity) and personal progress in that my home makes me feel I am doing well in life (self-esteem).

Research on housing and well-being has shown that dwellers adapt to their circumstances over time (Clapham et al., 2017). For example, a household moving to a larger house may initially experience an increase in well-being after the move, but this reduces as they become accustomed to the new situation (Foye et al., 2017). Housing is, at least in part a positional good (Hirsch, 1977), and so its impact is influenced by factors related to social status. A dweller's perception of social status may be influenced by who they compare themselves with, in other words, their reference group, but little is known about this (Foye et al., 2017).

There has been much interest on the importance of tenure as a factor in the achievement of home. As we saw in the previous chapter, tenure structures the rights and obligations of residents and owners in relation to the house. Tenure forms are created by governments through legislation and can be amended through the legal system – such as by legal precedent in the courts. But legal rights and obligations are largely important to the extent that they become embedded in social practices that will define the experience of residents. Tenure forms also have cultural dimensions and can vary in the individualism or collectiveness of their provisions as we saw in Case Study 4.3 of Swedish Housing Co-operatives. Two points are important from this discussion for our purposes here. The first is that the concept of home has cultural dimensions that influence the social practices. For some people, home is experienced as an individual household phenomenon based around the individual or the nuclear family. For others, home has elements that include the co-operative responsibility for enjoyment of common parts of houses and communal residential social practices as shown in the concept of co-housing that is evident in Sweden and other Scandinavian countries in particular. In some schemes, common household tasks such as the making and eating of food or cleaning may be shared and undertaken communally. In these cases, the concept of home has a substantial collective element.

The second point is that different tenures influence the social practices in a house that may influence the ability of residents to feel 'at home'. This may be for two sets of reasons. The first relates to the social practices that tenures enable or inhibit. For example, a private sector tenant may not be able to decorate their own house and may have less security of tenure than residents in other tenures, and this may influence their feeling of being at home. The second point is that tenures will vary in their symbolism and status. For example, there has been much research in the UK on the impact of owner-occupation on feelings of home. Gurney (1999) has written about the normalising discourse of 'home ownership' in government documents that equates the concept of home with that tenure. The discourse assumes that people own homes and rent houses. The discourse of 'home ownership' is reinforced in everyday discourse through aphorisms and other common forms of speech.

In summary, a house has many affordances. Some relate to the social practices of everyday life such as eating and washing that are carried out in the home. Others relate to the meanings that home has for us and for others. Both these elements are related to the physical design of the house but are not reducible to it. Meanings can be reinforced or changed through use and are influenced by wider social processes and discourses. There are two main consequences of this is for our discussion of housing quality. First, as outlined in Chapter 4, a house is a complex phenomenon that is different from other commodities in a way that makes orthodox economic analysis difficult. As a result, housing markets do not behave in the way that neoliberal ideology and neo-classical economics would predict. The nature of the home discussed here makes this picture even more complex. It is not just as an economic commodity that a house is a difficult phenomenon to analyse. As we shall see, it is a difficult subject for housing policy also. The second consequence is that the affordances of a house relate to a specific individual and may be very different for individuals. Again, the implication of this is that the question of what is a good house is a very personal one that is difficult to generalise about and so makes policy-making in this area difficult to pursue.

Nevertheless, it is not an area that government can ignore as the review above shows just how important the achievement of a feeling of being at home is to a dweller's well-being.

Well-being itself is of course a very subjective concept, but it offers a good way of moving beyond the policy concern with the physical fabric of a dwelling and on to the impact that the affordances of the house have on the well-being of its occupants.

Reasons for government intervention in the quality of housing

The key questions to be addressed by policy-makers are why and how should government intervene in the housing market by making policy on the quality of housing. There a number of justifications for government action to improve the quality of housing. Government may act to safeguard the residents and builders of housing against conditions such as noxious or poisonous substances, on the grounds that, otherwise, the well-being of the dwellers would be harmed, and it possesses the technical knowledge that may not be generally known or accessible to market agents. Intervention may be justified if there is market failure in the sense that some households cannot purchase or rent in the market a standard of housing that is deemed to be acceptable. Left to itself the market outcome would be deemed to be harmful to the residents or the population at large. These two justifications depend on government being concerned with harm to the dwellers themselves. Another justification may be recognition that there is an imbalance in power between residents and developers or landlords that requires the state to step in to rebalance the relationship by enforcing measures in favour of residents to enable them to be able to enjoy appropriate housing. In other words, it is accepted that the market does not function well in responding to all demands placed on it. A further justification is that the quality of homes has externalities that are not considered in a market context. For example, a house that is not structurally sound may endanger passers-by as well as those who live in the house. If a house is conducive to infectious disease, then this may lead to a widespread epidemic that harms many other people as well as the resident. But there may also be externalities that relate to social objectives that are not taken into account in market transactions. For example, there may be externalities in terms of environmental sustainability. Thus, it may be considered to be in the general interest that houses are built to standards that minimise energy usage. In this case there is deemed to be an overall societal interest that is not accounted for in the market interactions and so needs government action to ensure it is achieved (see Chapter 11).

Therefore, the justifications for government action rest on three factors: the incidence of harm to dwellers and others; externalities where the costs and benefits of individual producers and consumers do not reflect a societal interest; or market failure that makes securing adequate housing difficult for some dwellers. The first of these may be consistent with a neoliberal position and is recognised in neo-classical economics, although there may be differences in the trigger levels for intervention and its extent between governments of different ideologies. However, the other two may be more contentious and may be considered more important by governments that see housing as a social right rather than as a marketed commodity. Each of these justifications will now be considered in turn and the evidence for them assessed.

Harm to dwellers and others

Studies have shown a relationship between poor housing and poor health, but there have been difficulties in showing a general, direct relationship between the two (for a review, see Clapham, 2005). In some particular instances, the relationship is clear because there is a strong causal link. For example, dampness leading to mould growth has been shown to be associated with respiratory problems, and there is a clear link between the two (see, for example, Marsh et al., 1999). Also, cold or excessive heat may cause hypothermia or overheating that may cause illness or death in extreme circumstances. Structural defects such as danger of collapse or noxious substances such as gas leaks or asbestos clearly can directly cause ill health. However, the evidence in other circumstances is less clear-cut. For example, there is evidence that, if people are rehoused from poor into better accommodation, their health improves (Byrne, 1986); however, it is unclear why this is. The key question is whether the most important element is the physical conditions themselves or an increase in self-esteem. The most plausible hypothesis is that the link between poor housing conditions and poor health is mediated by social psychological factors such as self-esteem, efficacy and social relatedness. It has been shown that poor housing conditions are associated with poor mental health (Clark and Kearns, 2012) that and poor mental health is associated with poor physical health. Halpern (1995) argues that the physical environment is mediated by social factors in its impact on the mental health of the individual and that there is an imperfect match between the immediate environment and the subsequent responses. Therefore, the symbolic meaning of an environmental stimulus is important in mediating the response. 'An objectively similar sound may lead to dramatically different affective responses, even with the same person if the context and appraisal differ' (Halpern, 1995, p. 38). Therefore, it is possible that government action to improve physical housing conditions may not improve the well-being of the dwellers if it leaves them feeling a lack of control and experiencing shame.

The possible mediating influence of mental health is important for housing policy because it has an impact on the success or otherwise of policy interventions. A focus on the subjective aspects of the relationship between housing conditions and health raises the possibility that what is harmful for one person may not be for another. Various factors may influence the outcome. For example, individual factors such as attitudes, expectations as well as psychological resilience may be important. All of these may be influenced by cultural factors. For example, there may be substantial differences between countries and over time about what is a socially acceptable dwelling. Conditions that threaten self-esteem and health in one country may not in another. Immigrants may bring with them expectations and attitudes from the country they are leaving that may be different to those in the country they are entering.

Also, a consequence of the importance of subjective factors is that improving physical housing conditions may not always or necessarily result in better health outcomes if the changes made do not influence self-esteem. In other words, the provision of better shelter may not by itself lead to improvements in health because shame in the way that dwellers are treated may outweigh any physical improvements. What is done and how it is done may be crucial in determining the outcome. An example here is the treatment of homeless people in the UK. In the SUSTAIN project (Smith et al., 2014), researchers examined the experience of homeless people who had been rehoused by local authorities into the private rented sector. It was found that few had been able to feel 'at home' in their situation because of poor relations with the landlord, poor physical conditions or insecurity of tenancy. Although people had shelter and their physical conditions were better than being homeless, the change was insufficient to create the feelings of home.

There are other factors in a house that may create harm. Standards in many countries include elements that relate to the amount of space in dwellings, the size of rooms or the number of people in a particular sized dwelling. The existence of overcrowding as a problem stems initially from concerns raised by laboratory studies of rats. In humans, as in rats, laboratory studies have shown that high social densities are unpleasant and anxiety provoking, attraction for others is reduced, helping behaviour is inhibited and both aggression and withdrawal increase (Halpern, 1995). However, the importance of space standards is contested, and the idea that there is a universal minimum space level below which harm is done to residents is questioned. Clearly there are cultural dimensions here as levels of space vary considerably between cultures and countries. Therefore, expectations of what is a reasonable space for family life are crucial to any setting of an appropriate standard. Changes in the average household sizes in many developed countries have led to arguments for the reduction in traditional space standards. For example, in the UK the move towards more young, single households led to the 'solo' with one room and bathroom being considered to be appropriate for this group. There are fears that overcrowding in terms of the ratio of number of people to the space in a dwelling can influence factors such as the ability of children to find the space and peace and quiet needed to do homework and that this can impact on their educational achievement. There are cultural values about the necessity of children of different genders having to share sleeping space, and this was an important element in Victorian overcrowding standards and is evident in the UK today, although it is also common in other cultures.

Halpern (1995) emphasises the importance of the feeling of being in control in the outcomes of overcrowding. In experiments, subjects who felt they had more control over the situation were less likely to experience high density as crowded. Also, Halpern (1995) points to the importance of a common group purpose as mediating the impact of high density. In other words, overcrowding is likely to be more acceptable within a family that has a shared life and objectives than with a household made up of people who are independent (an example may be sharing in a homelessness hostel). Therefore, there is a relationship between the density within a dwelling and mental health, but it is not invariant and does not hold across different settings and cultures. The findings of general studies are made complex by the poor mental health of those living alone. In shared accommodation (with non-kin), it is the large number of unpredictable and unwanted encounters with acquaintances and strangers that lead to the experience of crowding (Halpern, 1995). The ability to regulate social interactions both within the home and the neighbourhood is important in fostering a feeling of control and avoiding feelings of helplessness.

Again, as in health arguments, there are no universal standards of space and overcrowding that can be shown to impact on residents. The affordances of houses are individual and are influenced strongly by individual experiences, attitudes and expectations and cultural norms and values.

Some harm may impact on other people as well as the dwellers. For example, a breach of structural integrity in a house could impact on others and not just the residents if the house were to collapse. The same argument can be applied to some health issues especially in areas where infectious diseases are still rife. Therefore, this area is a common one for government intervention, and it can be argued that some of the first interventions in housing through public health legislation were made under this justification.

The subjectivity of harm factors makes government intervention difficult to design and makes its impact on well-being difficult to predict. Some interpretations of the neoliberal discourse would adopt a *caveat emptor* (buyer beware) view of housing standards. The assumption is made that people are the best judges of their own welfare or well-being and they should be left to make their own market choices. In a competitive market, it is assumed that supply will be responsive to the demands of consumers. An alternative social-democratic view, based on the idea of housing as a social right, may be that there are information asymmetries and power differences in the market that justify intervention on behalf of the consumer to ensure that all citizens are able to enjoy a societally acceptable standard of housing. Intervention to achieve minimum standards is common in many countries in regard to consumer goods in general, where there is the potential of physical harm to consumers. In practice, many governments seem to accept the justification in housing, but to different degrees. There is variation in the factors that are considered important and in the standards that are required and levels of enforcement may differ.

A societal interest

The arguments for state intervention in the quality of houses with regard to externalities are complex, and there are two important elements to the externalities argument apart from the harm element considered above. The first is that houses last a long time and so the present residents may not be the only residents. A government concerned about best use of the housing stock may want to influence standards to enable future residents to enjoy the affordances of home as expectations increase. Market pressures may not lead to a satisfactory outcome in this. For example, a developer of owner-occupied houses may only be interested in the first sale if this is sufficient to ensure a profitable sale and high share value. The first purchaser may be subject to affordability constraints and may only be planning to live in the dwelling for a short period of time. A rational consumer would be concerned about the future saleability of the dwelling, but not all consumers are rational, and their timescale may be short. There may also be differential opportunities to adapt and improve a house in the future due to their form and design. One reason for the continuing success of Victorian terraced houses in the UK has been their flexibility that has enabled residents to alter the internal design to accommodate changing tastes and aspirations. Governments may take the view that they are the best judges of the long-term interest of consumers in order to ensure the effective working of the housing market in future, either by insisting on higher standards that would be more applicable when preferences can be expected to increase over time, or by ensuring the flexibility that would enable dwellings to be more easily improved and adapted to meet changes in demand.

The second externalities argument is that there may be wider societal objectives that individual producers and consumers may not share. In other words, as we argued in Chapter 4, individual consumer interests may not be aligned with social objectives. The most common of these arguments is over environmental sustainability objectives, which will be examined in more detail in Chapter 11. An important issue is the energy efficiency of homes and the general environmental impact of their construction and use. Profit-making housing developers and builders have little interest in building to high environmental standards if they can sell lower-standard properties at a higher profit. Therefore, their attitude is likely to be driven by consumer demand, although the power imbalances between producers and sellers could inhibit these being met if producers do not see it in their interests. It could be argued that consumers have a strong interest in their properties being energy-efficient as it will result in lower costs for them. However, consumers may not be aware of the issues concerned, and there is likely to be a financial trade-off if low running costs are the result of higher production costs that mean a higher purchase price.

The uncertainties over the actions of the consumers and producers and the lack of meaningful action have seen many governments taking the lead by imposing housing standards that lead to improved environmental standards of construction and use. Examples are the move towards carbon-neutral housing in the UK (see Chapter 11). The argument for state involvement is based on the view that governments are better placed to have a view of the interests of the society of a whole and for future generations. In addition, as we will see in Chapter 11, environmental sustainability issues are complex and wide ranging. For example, meeting international objectives on mitigating climate change will involve action in a large number of different fields and governments are in the best position to co-ordinate this. They are also best placed to put pressure on housing producers to change behaviour if the market is not sensitive to consumer demands.

Market failure

It is because markets fail that people are harmed and externalities are not considered. However, in this section, the focus is on situations where it is considered that some households do not have sufficient purchasing power to achieve an appropriate level of housing quality or the market operates in such a way that there are power imbalances between providers and residents that mean that consumers have little choice but to pay large sums of money for poor standard properties. This situation could exist where there is an overall shortage of property so that competition forces out the poorest or most vulnerable and the supply response is not forthcoming. But even in a situation of overall surplus, there may be locations where there is a shortage because of historical factors or uneven economic growth. It may be that the incomes of the poorest are not sufficient to purchase housing of an acceptable quality, although this is not a market failure as such and is considered in more detail in Chapter 8.

Governments have reacted to perceived market failures of this kind in the private rental sector more than in ownership. This may be because in many countries the poorest conditions are in this sector (although not all as we shall see later) and because the social relationships in rented housing are considered to be more imbalanced and with more severe consequences than in owner-occupation. It is in rented housing that the power relations between the provider or landlord and the resident can be considered to be more acute because of the additional responsibilities of the landlord and owner and manager of the property. This gives landlords an influence over the day-to-day life of a tenant that is different from that in owner-occupation where other agents such as banks do not have the same degree of influence over the day-to-day practices.

Although all elements of the lowest rungs of the private rented sector are subject to intervention from some governments, there is an additional focus on houses in multiple occupation where residents do not have self-contained accommodation because this is where house conditions are often at their worst and where the poorest sections of the population live.

Another important example of market failure relates to individual disability. In an unfettered market situation, disabled people would be forced to act as individual consumers to purchase their particular requirements. There may be a number of problems with this situation. First, some people may not have the income to be able to pay for any extra physical requirements. Second, disabled people may be a small percentage of the population and so not be able to effectively transmit their demands in a way that producers find worthwhile to respond to. Third, general standards of new construction may be such that costs of responding to the needs of disabled people may be high. For example, if every house does not have easy access or enough space to enable wheelchair use, then the possibility of renovation may be low and costs high. Fourth, the quality of life for disabled people can be enhanced by the ability to access buildings other than their own dwelling. This may relate particularly to public buildings, but also to the houses of friends and relatives. Therefore, a case can be made for the imposition of 'visitability' standards that afford access for some disabled people.

Dilemmas of government policy

The physical condition of a house may have an influence on health, but many other factors may be involved, and the relationship is unlikely to be a direct one. So how can a good-quality house be defined in a way that is useful in designing policy interventions? Clearly there are minimal levels of shelter and physical integrity that are required for health, although they will vary according to climate and other factors such as susceptibility to earthquakes or other weather events. If one adds into the picture the importance of cultural factors in defining the appropriate type and quality of house required there seems to be little room for universal dwelling standards. Any standard needs to be nationally and culturally specific.

One factor that may inhibit governments from intervention is the impact on the cost of housing. For new-built housing, the insistence on higher standards may increase the costs of developers and builders and so, in most cases, increase the price of housing. Whether the additional costs are borne by the producer or the consumer in the form of an increased price, or are passed on in lower land prices, will depend on market conditions as well as institutional structures. Where there is a shortage of housing it is likely that producers can pass on any cost increase to consumers. Where the market is less buoyant, it is possible that producers may have to bear some of the increase in cost in a reduced profit margin which will lead them to be able to afford less for their land. Of course, there may be knock on effects. An increased price may result in affordability problems for some households that may reduce demand and so possibly reduce supply. A reduced return to developers may also result in reduced supply as marginal developments become unprofitable.

In renting, an improvement in house conditions is likely to result in an increase in costs for the landlord, and again these may get passed on to the resident in the form of higher rental costs. Any increase may reduce the residual income of residents and so could create problems of affordability or of the ability to meet other living costs. However, if there is a housing allowance programme (see Chapter 8), the increase in costs may be passed on to the government.

The impact of housing improvements on costs could put governments in many dilemmas in relation to the outcomes of intervention, especially considering the subjective nature of the well-being afforded by a home. For example, a large group of new immigrants may share an apartment and split the cost between them, experiencing housing conditions judged by them as being satisfactory in the light of the conditions they have been used to in the country they have just left and the probable short time they will be living there. Does government intervene through overcrowding legislation to force the residents to live at lower density and higher cost? What justification is there for over-riding the choices of the residents and their own definition of what meets their subjective well-being? One possible motivation for intervention in the private rented sector is the power imbalance between landlord and resident. Therefore, should the appropriate intervention be to prevent the overcrowding or just to ensure that the rent is set at a level that prevents undue profit-making thus helping the residents? How can undue profit (or profiteering) be identified and measured?

Given the complex and subjective nature of the issues over housing quality there are many difficulties for governments seeking to intervene in this area. Nevertheless, most countries have adopted the concept of minimal standards in housing that relate primarily to the physical condition of the property. These are usually focused around the specific harm producing elements discussed above, but in some housing regimes, may also relate to the externalities.

Policy tools

Governments have a number of options for dealing with poor housing conditions. Many general housing market interventions will influence the condition of housing and the more efficiently and effectively the market works the less likely there is to be poor conditions. In addition, housing subsidies and general interventions to reduce income and wealth inequalities or to boost the incomes of the poorest would help. General economic conditions such as a reduction in unemployment could also result in a decrease in the incidence of poor conditions. However, here we will focus on the measures that could be aimed specifically at the quality problem and these mainly consist of regulation of the market in terms of the categories of government intervention identified in Chapter 2.

A very common intervention is the *regulation of new building* by the imposition of minimum physical standards. Most countries have some form of building code that covers all tenures, although there may be different standards for different sectors. For example, in the UK for many years legally specified minimum space standards in the public rented sector were higher than those in the private sector. The level of standards set varies considerably between countries. The advantages of this approach are the relative ease and cost-effectiveness of including standards at the time of building rather than attempting to improve standards later. Given the longstanding nature of housing, this may be important in the longer run. One disadvantage of this approach is that it will take time to impact on the housing stock as a whole because of the small proportion of the newly built stock in the total housing stock. Another disadvantage of the approach is that it adds to costs as discussed above. Clearly the extent of the impact will vary according to the level set and the scope of the standards. As discussed earlier, they may only include harm elements or may include externalities such as environmental standards.

Government could stipulate *minimum standards for the existing stock* and encourage or force existing owners to achieve them. It could do this through the criminal law by making the sale or letting of such property illegal and could institute sanctions such as fines or even imprisonment to achieve this. Alternatively, residents could be given the right to challenge inadequate conditions through mechanisms such as the civil courts, housing tribunals or by reference to professional inspection organisations. An alternative or addition to the use of sanctions could be the provision of financial incentives to owners or landlords to improve the condition of properties through the provision of financial help or subsidies through the provision of grants or loans. These can be used to overcome the 'valuation gap' that exists if the increase in value of the property after repair or improvement does not cover the costs involved. Help in this way is often means tested so it only goes to the poorest people who are deemed not able to afford it by themselves. Financial help can be for repair of property or for the installation of basic amenities.

Intervention over housing conditions has often focused on the private rental sector as it is here that physical housing conditions tend to be worse. One policy mechanism here is the licensing of private rented tenancies, where landlords have to register their property for letting and to meet standards laid down. Licensing may be comprehensive and wide in scope or limited to specific locations or types of properties such as Houses in Multiple Occupation (HMOs) or rooming houses. (See Case Study 5.2.)

Case Study 5.2 Rooming house regulation in New South Wales and Victoria

A traditional view of regulation is that it involves two parties: regulators and regulatees. Regulations are rules or directives that require regulatees (whether individuals or organisations) to act in particular ways and, more generally, aim to shape their behaviour. A more modern view is that regulation goes beyond a formal or functional account of government and rules to be observed by regulatees. A 'decentred' definition of regulation recognises a variety of economic and civil society actors who are involved in regulation through information gathering, setting standards and in shaping behaviours. This view recognises the economic and social processes that accompany all formal legally-based systems of regulation.

There are two important dimensions to decentred institutional arrangements: diverse actor group membership, and multi-level governance arrangements.

1   Many people and organisations are involved in defining goals of regulation and setting rules and standards as well as the on-going process of checking compliance with regulations. They also attempt in various ways to change behaviours to achieve the goals that underpin the regulations, for example through enforcement strategies, information dissemination and training.
2   Different agencies from all three levels of government in Australia's federal system are participants in the regime of regulation. As in many areas of shared responsibility in the federal system, there are different objectives and limited means for resolving differences in approach. Non-state actors are also actively involved, for example in representing the interests of rooming house owners and operators and rooming house residents.

Much research into regulation argues that regulation should be based on assessments of risk and take into account the motivation of regulatees and their capacity to comply with regulations as well as the resources available to regulators. It also identifies three types of regulation: prescriptive, system-based and performance-based regulation, the latter being concerned with outcomes.

Renewal of NSW and Victorian rooming house regulatory regimes

Rooming house regulatory regimes in NSW and Victoria were changed in 2012 through the Boarding Houses Act 2012 in NSW and amendments to the Residential Tenancies Act 1997 and the Public Health and Wellbeing Act 2008 in Victoria.

Both states retained decentred, multi-level regulation in which multiple actors were involved. The regulatory regime for rooming houses in NSW and Victoria, both before and after legislative changes in 2012, is prescriptive; it involves detailed rules on what regulatees must do and how they must do it.

At the same time, regulation in response to a rights-based agenda driven by the Australian Government sought to extend access for people with disabilities to single room accommodation in rooming houses.

Recognition of risk

The major risk to residents in rooming houses was seen as relating to fire hazards. A (bare) majority of Investigative Panel members across the two states supported the view that new state-wide registers had improved rooming house regulation. In both states, there was said to be a considerable number of unregistered and therefore unregulated rooming houses.

Current regulation does not address other types of risks such as exposure to being the victim of criminal activities and experiencing behaviours of other residents consequent to various mental health conditions. These behaviours were thought to be associated with use of rooming houses by government agencies for discharging people from institutions of various types and require approaches other than prescriptive regulation.

Accountability and incentives for effective regulation

Issues of accountability and incentives for effective regulation come to the fore within a context of rather diffusely distributed regulatory responsibilities in the post-2012 environment in both states. Some of the concerns/issues expressed in the research were the lack of incentives for local government to take a more proactive approach to unregistered rooming houses and whether local governments had been given the resources to effectively take on additional regulatory responsibilities.

A risk-based approach to regulation is required which distinguishes between the risks associated with buildings and the risks associated with housing vulnerable people. Some segments of the market may be lower risk and market factors may be more effective than prescriptive regulation. Others carry higher risk and may require additional regulation as well as improved inspection and compliance. Some of the risks associated with vulnerable residents may require service interventions rather than a regulatory approach. Clearer accountability for regulation is important as are financial incentives for effective regulation.

It is important to develop clear objectives for regulation across the entire system, which would enable regular assessment of the performance of the regulatory regime. The costs of regulation should be explicit for different actors and consideration given to how costs and income could be distributed to provide incentives for better regulation.

Current regulation focuses on the safety and amenity of buildings. Options for improving rooming house operation include education/training (as in Victoria) and/or licensing of operators. This would bring rooming houses in line with other areas such as child care and aged care where there are vulnerable clients.

This case study shows the complexity of regulations systems and the different ways that regulation may be used.

Source: Based on Dalton et al., (2015)

Case Study 5.2 shows that regulation is not a simple issue with a variety of agents and relationships involved. Regulation may be formal, being based on detailed legal requirements, or more informal and based on agreed compliance of agencies involved. Appropriate incentives are needed if the objectives are to be achieved and a deployment of financial and other resources. Regulators need to have the appropriate resources and knowledge to undertake their task.

The example of Grenfell Tower in the UK (Case Study 5.3) shows that regulation can be a problematic activity, particularly where there are complex accountability arrangements instituted through neoliberal mechanisms of market and quasi-market relationships accompanied by severe public budgetary restrictions. This case study shows how the objectives of regulation can be subverted by the agencies involved if it suits their organisational objectives. In this case, general regulations were reinterpreted when specified into enforceable form by professional bodies and private interests that had different objectives than government.

Case Study 5.3 Grenfell Tower – UK

In June 2017 Grenfell Tower, a 34-storey high-rise block of apartments in Kensington, London, caught fire. The fire was started by a faulty fridge/freezer in one apartment, but quickly spread through the whole building. Although the fire-brigade was quickly on the scene, they struggled to control the fire, and there was a very high loss of life with almost 100 people being certified as dead or missing.

The block was originally designed on the principle that fire would be contained within each individual apartment, and there was a stairwell that should have provided a safe means of exit to all residents. In this case the fire spread from one apartment to the entire block by means of external cladding that had been added a few years previously in order to increase insulation for the apartments as part of the move to meet sustainability targets, as well as to improve the external appearance of the building.

In the aftermath of the fire a number of questions were asked about why it could occur that shed light on the nature of the building regulations. One focus was on whether the cladding panels that had helped to spread the blaze and which, after tests proved to be flammable, had been allowed under national building regulations. Answers to this question seemed to vary. The government claimed that the panels did not meet current regulations and should not have been used on a high-rise building. However, others argued that the government had been watering down building regulations in the previous few years in their attempt to reduce 'red-tape' and had ignored the findings from previous fires in high-rise blocks where calls were made to install sprinkler systems in blocks and to review the building regulations in relation to fires. It was discovered that government had attempted to reduce the fire requirements in the building regulations for new schools in order to reduce costs.

The building regulations were argued to be vague and not kept up-to-date, and so building surveyors tended to rely on professional and commercial guidance that 'interpreted' the statutory guidance and in this case appears to have downgraded the fire requirements based on 'lap-top' studies of combustibility.

A further question arose about the enforcement of building regulations which is in the hands of building inspectors employed or controlled by local authorities. It transpired that the cladding of the block had been inspected on site sixteen times, and so questions were asked about why the unsuitability of the cladding had not been discovered. In mitigation, it was argued that funding for building control inspections had been reduced in line with cuts to total local government expenditure enforced by central government. Financial cuts to fire services were also blamed by some for the lack of appropriate fire advice to developers and of fire inspections of new buildings.

The residents of the high-rise block were furious that their pleas for action before the fire were ignored by the local council and their landlord which was a private company to which the functions of the council had been contracted out. They had campaigned, without success, for fire safety measures after a previous fire in a high-rise block, Lakanal House, a few years before.

Generic issues:

- Trade-offs between different reasons for regulations, e.g., between fire safety and sustainability goals
- Trade-offs between requirements and costs to developers or landlords
- Need for enforcement of regulations and the organisational systems and resources required to achieve this.
- The number of interests that are involved in the formulation of regulations such as fire services, building surveyors, politicians and local residents with varying degrees of power to achieve their desired solution.

As well as these specific mechanisms, the existence of unsatisfactory housing is a consequence of the functioning of the housing market and wider societal factors such as inequality of wealth and income, and so measures that impact on housing inequality in general will have an impact on house conditions. Many of the policy tools to achieve this will be considered elsewhere in the book.

Evaluating interventions in housing quality

In the book as a whole we have taken our evaluation criterion for housing policy as the well-being of individual dwellers, and the subjective nature of housing quality means that this approach is appropriate here. However, there are additional criteria, which are the needs of future generations and sustainability issues, which will be considered further in Chapter 11.

The subjective nature of well-being derived from a house means that it is difficult to assess the impact as changes in physical housing conditions may have a very different impact on individuals and so can only be judged on this basis. Subjective well-being from the house is, also, influenced by the general social context. As we have seen, the minimal standards set by government can vary over time and place and are essentially based on societal expectations that also influence individual expectations. Dwellers base their judgements of their position on the situation of reference groups. As we noted earlier, there is little known about who dwellers compare themselves with or what influences this. However, it is likely that the degree of inequality in housing circumstances and the perceived validity of this will be important factors. General well-being is higher in more equal societies, and there is some evidence reviewed in Chapter 4 that the same pattern is true for well-being from housing.

Housing regimes may differ in housing quality in two ways. First the type of regime may alter the incidence of poor standards and the relative positions of households. Housing regimes where there is extensive income inequality or low economic development and high market penetration are more likely to have poor housing conditions in the lower strata of the housing sector. Although low overall economic level is likely to result in low absolute physical housing conditions, a high degree of income and wealth inequality will result in greater differences between the housing conditions of different groups, creating poorer relative conditions. The positional nature of housing and the importance of self-esteem in well-being mean that relative standards may be important in dweller well-being. In terms of housing outcomes, countries may differ in the extent of inadequate housing, influenced by the structure and the effectiveness of the house building industry (see Chapter 7). Housing supply systems differ in their sensitivity to changes in demand and to different segments of housing demand.

The institutional structure of the housing regime will influence the desirability of intervention. This particularly applies to the rental sector where much intervention is focused. Where private landlords are large institutions with a business plan that is based on long-term returns, it may be that higher standards are more acceptable and have fewer disadvantages. The key point is that outcomes will depend on the specific institutional structure in an individual country at a particular point in time and therefore there are limits to the ability to generalise.

Second, the housing regime may influence the likelihood of government intervention to improve standards. One would expect governments with neoliberal ideologies to be less likely to want to intervene as they would be more likely to accept market outcomes. Nevertheless, neoliberal governments may intervene to reinforce a market ideology by removing the worst outcomes and thus helping to avoid protest and illegitimacy of the market. The concentration on minimum standards may help to overcome some of the problems involved in the subjective nature of the affordances of housing. However, governments may differ in the extent to which they are prepared to set minimum standards at a level that increases standards above where they would be in the market. Governments may also vary in the scope of the standards they set. For example, some governments may relate the standards to new-built housing or others may also include existing housing. It may be that the imposition of standards is only applicable in some locations and not others. It is possible to set standards in a way that just prohibits factors that would cause individual harm as outlined above and ignore externalities. Other governments may wish to go further than this and use intervention to drive increases in standards, an approach that has been used with reference to sustainability goals (see Chapter 11).

Conclusion

The chapter has focused on the house and home, using the affordance approach to analyse the utility and meaning of a house for dwellers and how these two elements are strongly interwoven and subjective. Apart from a few defined situations involving clear health triggers, the well-being of dwellers is personal and related to the meaning and status aspects of a house. There are a number of justifications for government involvement in the quality of housing that relate to the harm caused to dwellers or various forms of externalities such as harm to others or the importance of wider social objectives such as sustainability and the needs of future generations. However, the subjective nature of the well-being derived from houses makes it difficult for governments to design appropriate tools and methods of intervention. The most common form of intervention is to impose minimum standards on new construction through building regulations. The problem is that standards are essentially arbitrary and subject to dispute between different interests as well as political ideologies and cultural expectations of what is acceptable in that society at that point of time. As a result, the propensity to intervene and the standards enforced vary considerably between countries as do the means of regulation that can vary in their complexity and effectiveness as the case studies showed.

There is an important contradiction at the heart of government intervention on this topic. The subjective nature of well-being derived from a house means that personal control is a key factor, and so it would seem to be best to allow individuals to make their own judgements about the standard of housing that best suits them. However, the differences in resources in many societies and the imperfections in the housing market mean that a consequence would be for some dwellers to live in unacceptable housing conditions from both their and society's point of view. The most effective way to attempt to deal with this contradiction would be to focus attention on inequalities in the housing market which, if successful, make intervention in house quality less vital. The focus can then be on externalities such as the needs of future generations and sustainability questions.

Chapter 6

Neighbourhood

Introduction

In Chapter 4 we noted that every house is unique because it has a unique location that brings with it access to public and private facilities as well as the social interactions with people who live close. The focus on the neighbourhood, or the locality in which a house is situated, is justified by its importance in the use and value of the home.

In some societies, it is common for houses to be 'open to the street' with permeable boundaries between house and neighbourhood, whereas in others there is a firm division between private and public space. Therefore, there is a relationship between cultural and social practices and the physical layout of houses and their localities. Also, neighbourhoods carry social cues about status and culturally valued lifestyles. The need to consider all of the elements of use and social meaning together was made in the previous chapter in relation to houses themselves, and it was argued that the concept of affordance was an appropriate tool to utilise in examining these factors, and so it will be used here for the discussion of neighbourhoods.

Housing policy has often focused on neighbourhoods. Some of the first interventions in housing were at the neighbourhood level, whether this was public health reforms or slum clearance programmes. There has often been a policy recognition that the neighbourhood strongly influences the well-being of dwellers there and that there are externalities in terms of crime or health concerns that spread beyond the immediate locality. Of course, in these cases it is usually poor neighbourhoods that are considered to be a problem and to warrant state intervention. Therefore, it is the concentration of low-income households in a particular location that is often the perceived problem that brings forth policies aimed at changing the nature of neighbourhoods whether through regeneration, gentrification or the creation of a 'balanced' community.

The aim of this chapter is to consider the affordances of neighbourhoods and to assess the policy problems and policies designed to influence their nature. The chapter begins with a definition of neighbourhood and an overview of the main attributes or affordances that impact on the life of residents. This is followed by a discussion of neighbourhood change, because their form is not static, but changes over time. An understanding of the forces for change is important in designing appropriate policies. The focus will then move on to policies towards residential neighbourhoods, focusing first on the way that problems are defined and the relationship to different housing regimes. This is followed by an examination of the main forms or mechanisms of government policy. Finally, the evidence on the impact of neighbourhood policies on the well-being of dwellers is examined.

Neighbourhoods

The aim of the analysis in this chapter is to examine the physical and social space around a dwelling that has a use and a meaning for the residents. Therefore, the concept of neighbourhood is used here rather than other possibilities, such as locality or place, for a number of reasons. Neighbourhood is a term that captures both the physical and social relationship with a dwelling, as well as having a meaning in everyday usage to people in many different cultures. The concept also resonates with a large tranche of literature that is important for our purposes here.

Neighbourhood may be a very useful concept, but it is one that is very difficult to specify and use in practice as there are many definitions (see Galster, 2012 for a review). Most definitions refer to a particular geographical area or a physical or geographical entity, although it is difficult to define the extent. It is commonly accepted that the boundaries cannot be defined in any objective sense, and so concepts such as symbolic or subjective boundaries are often used. For example, Galster (2012) argues for the importance of perceived boundaries of residents (and others) in defining neighbourhood rather than any objective measure. He also follows Suttles (1972) in conceiving of a multi-level spatial view of the perception of neighbourhood boundaries depending on the particular attribute being considered. Pebley and Vaiana (2002) use three classifications of levels of perceived boundaries of neighbourhood. The first and smallest is the area in which children can be left to play without supervision; the second level is labelled the 'defended neighbourhood' which is the smallest area that is perceived as having a meaning in opposition or in contrast to another area. The third and highest level is usually congruent with local government boundaries and is a segment of the city in which social participation is selective and voluntary.

Galster (2012) sees a neighbourhood as a cluster of attributes that may have different perceived spatial boundaries. He conceives of individuals as having a perceptual 'externality space', which is defined as the area over which changes in neighbourhood attributes are perceived as altering the well-being the individual derives from the particular location. These externality spaces may be congruent with particular geographical boundaries or not, may be general in the sense that the spaces for different attributes correspond or may or may not correspond with the externality spaces of others.

The concept of a neighbourhood consisting of a number of attributes that may have different externality spaces for individuals is a very useful one. Using the concept of affordance outlined earlier these attributes can be seen as affordances of neighbourhoods that may (or may not) impact on the well-being of individuals.

Galster's (2012) attributes are as follows:

- Physical characteristics of the buildings, for example, design, density, state of repair etc.
- Infrastructure characteristics, for example, roads, pavements utilities, streetscaping etc.
- Demographic characteristics of the resident population, for example, age, family composition, ethnic mix etc.
- Class status characteristics, for example, income, occupation, education etc.
- Public service characteristics, for example, quality of public schools, policing, parks, health facilities etc.
- Environmental characteristics, for example, extent of noise, air or other pollution, views, topography etc.

- Proximity characteristics, access to major destinations of employment, entertainment, shopping etc.
- Political characteristics, for example, organisation and influence of local residents.
- Social-interactive characteristics, for example, extent and quality of social interactions, perceived commonality or community, extent of local voluntary organisations, strength of socialisation or social control forces (also sentimental characteristics; such as residents' sense of identification with place, historical significance of buildings etc.).

The list is an interesting one and a good starting point for our discussion here. However, its main drawback is the clear distinction made between the 'objective' categories and the one all-inclusive 'subjective' or sentimental characteristics that are only mentioned as a subcategory at the end. In our discussion of affordance, we have made the point that people perceive the physical and meaning elements of phenomena together. Therefore, it is more useful to see each of the above categories as having both physical and meaning elements that are indivisible. Thus, the physical characteristics of buildings in the first category also contribute to the meaning that individuals ascribe to the place and their identification with it, and similarly for each of the other categories.

In summary, the concept of neighbourhood alerts us to a number of attributes or affordances that may have different spatial dimensions, but which contribute to varying degrees, and in different ways, to the well-being of individual dwellers in the neighbourhood. The relative impact of the different affordances will depend on a number of factors such as the age, family circumstances, income, physical capacity and lifestyle of the individual. Therefore, it is possible in theory for a particular neighbourhood to afford high well-being for one person and low well-being for another depending on their particular circumstances. For example, proximity to facilities may vary on the basis of access to a car or other means of transport or the ability to drive. The use of some facilities may be made physically difficult or emotionally demeaning because of physical barriers to entry or humiliating procedures of entry for people who have a physical impairment.

Neighbourhood social interaction and community

There has been much written about the perceived 'decline in community' (for a review, see Clapham, 2005). The argument is that there existed a traditional working-class community built around overlapping ties of family, friends and neighbours that flourished in particular neighbourhoods in Britain (such as Bethnal Green) at particular times (the 1950s) (see Young and Willmott, 1957). Since the 1980s, this particular form of neighbourhood interaction seems to have been an anomaly rather than the norm, as numerous studies have not found this pattern in other places and at later times (see, for example, Holme, 1985). There is a strong argument that community in this form is unlikely to exist to any great extent in the future as physical and social mobility has dislocated ties of family and friends from those of neighbourhood. Many people are increasingly mobile through the use of cars and other forms of transport and can pursue hobbies and leisure activities and maintain friendships over a wide geographical area. In addition, new technologies have meant that shopping and leisure activities and social networks can increasingly be pursued at home through the internet thus obviating the need for physical mobility.

Local neighbourhood relations are now characterised as 'weak or loose ties' (Granovetter, 1973; Henning and Lieberg, 1996). Nevertheless, Henning and Lieberg (1996, p. 22) argue that the value of these weak ties should not be underestimated. 'The significance of loose ties was underlined by the inhabitants who stated that these contacts meant a "feeling of home", "security", and "practical as well as social support".

Henning and Lieberg (1996) stress that these weak neighbourhood ties are particularly important to vulnerable and marginal groups who are likely to be less geographically mobile than others. Willmott (1986) argues that there is an ebb and flow of interaction with neighbours over the life-cycle. High interaction tends to occur at stages when more of life is undertaken within the neighbourhood such as in childhood, early parenthood and old age. The perceived decline in community could be due in part to the decline in the number of households in these categories with the increase in single person and childless couple households. In addition, the increased employment of women and increased access to a car also reduces the number of those who live out a large proportion of their lives in the neighbourhood. One result of these trends is the growing exclusion of those who are limited to the neighbourhood from those who live most of their lives elsewhere.

The weak ties that characterise many neighbourhoods could lead to the conclusion that the locality does not matter in many people's lives. There has been much emphasis on the globalising influences that have meant that people increasingly are not tied to localities in their everyday functioning. Shopping and socialising can be undertaken on the internet, and modes of physical mobility mean that neighbourhoods are driven through in a car rather than actively used. Nevertheless, there has been increasing interest in the way that people relate to their neighbourhood despite these globalising tendencies. Savage et al. (2005) draw attention to the way that middle-class people in Britain associate themselves with their place of residence in a process they call 'elective belonging'. People want to feel that they belong to a place and weave it into their biographies in an active process of identity formation. People relate the places they live in to their lives and construct their personal identity in part around their relationship with the place. They do not have to use the place in any physical way, but it has meaning for them that is a constituent of identity formation. As Forrest and Kearns (2001, p. 2130) put it 'in a sense the neighbourhood becomes an extension of the home for social purposes and hence extremely important in identity terms: "location matters" and the neighbourhood becomes part of our statement about who we are'.

In the views of Giddens and other late modern or post-modern writers, neighbourhood becomes an important counterpoint in the lives of many to the insecurities of a globalised world. Scase (1999, p. 54) argues that:

> In an uncertain world where jobs are insecure and futures are unpredictable, living in a risk society reinforces the importance of community in a symbolic sense. Individuals obtain a sense of 'place', of attachment to the communities in which they live.

Clark and Kearns (2012) examined the psychosocial environment, which they defined as the nature of social interactions within the neighbourhood context and how these interactions made people think, feel and behave. They argued that a good psychosocial environment promotes a positive experience or view of oneself in relation to others, for example, in terms of control, confidence, self-esteem and status. In their study of deprived areas of Glasgow, they found a range of neighbourhood factors associated with high mental well-being. These included the perception of attractive buildings and environment in the area, feeling safe walking at night, community influence on decisions and a sense of personal progress (in other words a positive view that the neighbourhood represents a good personal progression). Dean and Hastings (2000) have shown the negative impact that shame and feelings of lack of control over the neighbourhood can have on the well-being of dwellers. As well as the perception of a neighbourhood by the people who live there, the perceptions of the media and nearby dwellers may, also, have an impact on well-being if that leads to a feeling of low social status and shame.

Neighbourhood dynamics

It is important to recognise that neighbourhoods are not static, but are in a constant state of flux. Understanding the processes of change in a particular neighbourhood are important in deciding whether and at what time intervention is necessary or desirable and designing the appropriate policy tool.

Scase (1999) predicted in 1999 that neighbourhoods would become increasingly segregated by demographics and lifestyles as well as income and occupation. Older people would continue to move to areas that match their lifestyles and provide facilities for hobbies, socialising and shopping. Young and middle-aged single households would gravitate to the inner city for the cosmopolitan and fashionable lifestyles based around leisure facilities of all kinds. Those with more traditional family patterns and lifestyles would live in the outer suburbs where provision of quality education and childcare would be concentrated. Other neighbourhoods would consist of those culturally and economically excluded who are constrained into locality-based lifestyles, dependent on the quality of local facilities. These trends are clearly apparent, although to varying extents, in many countries, including the ones considered in this book.

Scase's analysis places emphasis on the choice element of residential location decisions. However, choices are made on the basis of available options, and these may vary according to the position of the household involved, whether this refers to their economic situation, employment, family structure, ethnicity or their physical and mental health and abilities. The dichotomy between choice and structural forces has been explored in the example of ethnic minority segregation. There is evidence that ethnic minority households tend to be concentrated in particular neighbourhoods, but is this because they choose to live together for cultural, lifestyle or family reasons, or are they forced to congregate together by discrimination or economic factors?

Musterd (2012) identifies four traditions of thought in the study of residential segregation. The first focuses on individual behaviour and the choices that households make within their economic constraints and the supply of dwellings available (see, for example, Clark, 2009). The second tradition focuses on the impact of major economic transformation processes such as globalisation on different spatial patterns (Sassen, 1991). A third perspective refers to institutional structures such as welfare state regimes and institutions that produce segregation (Musterd and Ostendorf, 1998). The fourth perspective focuses on the place, time and context specific factors that create variations in places over time. The most useful way forward is to use an integrated perspective that incorporates all these levels of analysis. The concept of a housing pathway (Clapham, 2005) achieves this by focusing on the decisions that households make and the context within which they make them. Using this perspective neighbourhood change becomes a complex interplay of household choice and the wider discourses and other factors that frame the choice.

The complex processes of neighbourhood change can lead to patterns of decline or, alternatively, of increasing social status and improvement. There has been considerable research on gentrification as a market-based process that results in population change in certain neighbourhoods with more affluent dwellers moving in and displacing poorer existing residents (see, for example, Atkinson and Bridge, 2005). Population change is often reflected in physical change to the dwellings, as well as in local facilities that may alter to reflect the needs and tastes of the new population. One outcome of these processes is an increase in house prices that may reinforce segregation by making the neighbourhood accessible only to higher-income dwellers.

Segregation of higher-income dwellers is not usually perceived to be a problem (except of course by those forced out of the neighbourhood), but segregation of one group increases the likelihood of increased segregation amongst others, for example, low-income dwellers. In other words, it makes it less likely that neighbourhoods are 'balanced' with a mix of different social class and lifestyle groups. Thus, the creation of high-income areas and the existence of low-income ones are linked and have to be seen as closely related phenomena. The gentrification of one neighbourhood may result in another becoming unpopular and experiencing an influx of displaced low-income dwellers with consequent impacts on house conditions, house prices and local facilities. The differences between segregated neighbourhoods are likely to be greater, the higher the level of inequality in the society, and large differences in income and wealth will be reflected in large differences in the physical condition and affordances of neighbourhoods.

Neighbourhood problems

Political interest in neighbourhood problems has tended to focus on the existence of segregation where people of a similar ethnic origin, economic class, or lifestyle category live together. Of course, most people live in such neighbourhoods, and segregation is generally only perceived to be a problem if there are cumulative effects on the dwellers in the area or externalities to the wider society. One example of this is concern about the separateness of some low-income neighbourhoods from the wider society that can result in a process of social exclusion. Another important issue is perceived to be the segregation of ethnic minority dwellers in particular neighbourhoods.

Musterd (2012) draws attention to the disputed nature of the concept of segregation. He follows Johnston et al. (2007) in drawing the distinction between two elements of segregation, namely separateness and location. Separateness concerns the degree to which members of a particular group live apart from the remainder of the population. Location concerns the degree to which members of the group are congregated into high-density or low-status and low-affordance neighbourhoods. The distinction between the two elements of segregation is important when examining its impact. Most of the attention in the literature on the impact of segregation focuses on ethnic minorities and, in particular, the element of separateness. The disadvantages to the ethnic minorities themselves of separateness are said to be stigmatisation and isolation from the opportunities and social networks of the wider society. Also, the concentration of economically disadvantaged people in a neighbourhood can lead to a shortage of facilities because of the overall lack of spending power. From the viewpoint of society, there has been concern that separateness can lead to a lack of integration and the continuance of a minority culture. The advantages of separateness for ethnic minority households may relate to the maintenance of social and family networks as well as the continuance of a particular culture. Separateness can facilitate the development and maintenance of specific cultural facilities such as religious buildings and specialist shops. It can also provide re-assurance and defence in a situation of discrimination. The mixing of people with different norms and values can lead to problems of conflicting lifestyles. As van Kempen and Bolt (2012, p. 450) note:

> [N]ot everyone likes garden parties with loud music at night and not everybody thinks that public squares are for children playing football and the bus stops are for hanging out. The use of alcohol in public may annoy Muslims, while Muslims wearing a burka may annoy natives. Confrontations between different groups are more common in areas with a mixed population than in mono-ethnic areas. Living closer together may mean that people are more aware of differences between themselves and others.

Interestingly, van Kempen and Bolt (2012) note evidence that members of minority ethnic groups living in areas that are almost entirely native have a higher chance of developing schizophrenia than minority groups living in areas of a concentration of ethnic minorities. Although the precise causal factors here have not been explored, it seems to offer evidence for the supportive impact of living next to people who are like ourselves.

The construction of the separateness of ethnic minorities as a problem depends on the political expectations of appropriateness that in turn are influenced by different views of the impact of cultural and racial mixing. For example, terms such as 'multi-culturalism' have been used with different meanings. The term can be constructed as recognising that people have the right to live differently and that these differences should be respected and people allowed to choose their own cultural beliefs and lifestyles, including choosing where to live. Alternatively, it may be asserted that there is a duty of ethnic minorities to integrate with the native population and to accept important cultural norms of the wider society. In the latter construction, separateness is perceived as a hindrance to integration.

The political concern with the existence of segregation amongst ethnic minorities has been coupled with a focus on the segregation of poor households. There has been a long-standing policy concern in many countries with the concentration of disadvantaged or poor people in particular neighbourhoods. The extent of this problem is dictated by the way that statistics are collected and analysed. The common use of small area statistics leads to a focus on those areas that have a higher proportion of deprived people than others. However, as Holterman (1975) pointed out for Britain, the majority of poor people do not live in poor neighbourhoods, and the majority of people in most poor neighbourhoods are not poor. Despite the clear evidence of the inefficiency of an area-based focus as a mechanism for reaching poor people, it has been adopted by many governments as one of the primary instruments of the alleviation of poverty, usually under the umbrella of the concept of social exclusion (or social inclusion). As Hulse et al. (2011, p. 5) write:

> There is increasing recognition of the role of home and place in contributing to social inclusion. Having a home involves not only having a roof over one's head but also provides a safe and private environment in which intimate relationships can be developed and children nurtured. A home is the base for the routines of daily life, from shopping and socialising to schooling and working. A home is connected to place through a physical dwelling, and place may be important for self-identity, attachment and a sense of belonging. More practically, place shapes access to transport, facilities,

jobs and services. Home and place thus provide a foundation for participation in social, economic, cultural and political life.

There are many definitions of social exclusion, but Hulse et al. (2011, p. 13) offer a useful one as follows:

> [S]ocial exclusion refers to current circumstances in which some people are marginalised and unable to live a full life for a variety of reasons that may include, but are not restricted to, a lack of material resources. These reasons include lack of family support, social isolation, ill health and disability, not having a home or living in unsafe or inadequate housing, low levels of education, and inability to get a job.

There has been some interesting research on the impact of living in a neighbourhood with a high concentration of poor people on their life chances. The results of this so called 'neighbourhood effects' strand of research seems to be that living together with other poor people has a small but measurable effect on life chances, all other things being equal (for a review, see van Kempen and Bolt, 2012).

Wacquant (2008) has drawn attention to the advanced marginality caused by the growing inequality in many developed societies. He identifies a marginal group (the precariat) who can no longer find security in a liberalised labour market where often wages are insufficient to alleviate poverty and have been abandoned by the cuts in the welfare state and are increasingly disconnected from trends in the macro-economy. He argues that increasing inequality is being reflected territorially, with marginal people being segregated in poor-quality neighbourhoods with a lack of community organisations and a cultural history to provide support. These neighbourhoods are 'symbolically spoiled zones' where people suffer 'territorial indignity'. The result is spatial alienation where people lose the humanised and culturally familiar place where they can feel 'at home'. These places are socially fragmented and symbolically splintered as individuals try to distance themselves from the source of their shame by disowning the area and their neighbours. Therefore, in such 'symbolically spoiled zones' subject to stigma and lack of social integration, it would be difficult to achieve well-being. For Wacquant, social exclusion is the result of a neoliberal housing regime that reinforces the impact of neoliberal policies of inequality and labour market insecurity.

Wacquant's analysis draws attention to the importance of the social construction of the problem and its link to different ideologies and regimes. Concern about the impact of social exclusion may vary according to different social constructions of the problem. For example, the discourse of social exclusion tends to construct the problem as being caused by a mix of personal behaviour and choices with structural forces such as public policy and life chances dependent on economic status. The concentration of socially excluded people in particular neighbourhoods is explained by the workings of market forces and neoliberal policies of inequality, rather than the result of inadequate personal behaviour. Alternatively, other constructions such as that of a 'culture of poverty' have located the cause of the problem in the cultural and behavioural norms of the poor people themselves and the contagion effect of living in close proximity with each other, resulting in a perceived lack of appropriate role models and the growth of different cultural and behavioural norms from the rest of society. Clearly the choice of an appropriate policy mechanism would vary considerably depending on the way the problem is constructed.

Reasons for government intervention in neighbourhoods

There may be strong reasons why governments may not want to intervene in neighbourhoods. If it is assumed that the processes of gentrification and decline are brought about by the housing market and the residential choices of individuals, then one neoliberal view may be that neighbourhoods merely reflect market outcomes and that market forces will correct them as in the case of poor neighbourhoods being gentrified. However, not all neighbourhoods will be gentrified, and neoliberal governments may want to intervene for a number of reasons. One may be to alleviate the externalities arising, such as perceived problems of crime or disease, spreading to other areas. In a similar way, segregation of ethnic minority households may be perceived to have externalities in the form of consequences for society as a whole in terms of health or crime or social solidarity. The focus here is on the benefit to the wider society and not necessarily to the dwellers in the neighbourhood. This is a crucial point that is exemplified in some of the case study examples of regeneration later in the chapter, where it is clear that the current residents of the area are not intended to be the primary beneficiaries of the programmes.

It is important to note that countries that have adopted more strongly a neoliberal regime, such as the USA and UK, have moved away from a policy focus on private sector neighbourhoods and towards the public sector. This move partly reflects the increasing concentration of the poorest households in this tenure coupled with the aim of privatising the sector. It is allied with an unwillingness to intervene in the private market in the belief that market forces should be left to deal with any problems, as well as probably a belief that, following neo-classical analysis, this will happen. When the UK government intervened to provide improvement grants to owner-occupiers and private landlords in the 1960s and 1970s, the aim was to overcome the perceived reluctance to improve a property because of the 'value gap' said to exist in some neighbourhoods between the amount spent on improvement and the resultant increase in market price or rental return. However, there was criticism that this gap did not exist in many instances and the state was subsidising behaviour that would have occurred anyway through market processes. The shift away from a focus on private and towards public neighbourhoods, also reflects a desire to reduce the size of public housing and to make profit from the redevelopment of often valuable sites, as well as a lack of funding for the maintenance and reconstruction of public housing. Case Study 6.1 gives an example of regeneration of a public housing neighbourhood in London where the beneficiaries have been those from outside the neighbourhood able to buy the new housing and developers who make profit, but existing residents are moved from their local neighbourhood and some forced to leave London in order to find alternative accommodation. Local authorities also benefited by receiving increased revenues from local taxes as well as profits from the development that they are able to divert to other housing areas.

Case Study 6.1   Heygate estate, Southwark, London

The Heygate estate was built in the 1970s in a design that has been described as 'brutalist' and was home to more than 3,000 people. Due to a range of physical design challenges, such as poor security, low energy efficiency and environmental issues, it was agreed by the council to rehouse residents and demolish the estate completely to make way for a regeneration of the area.

The rehousing of residents began in 2008, and the estate was first fully secured in preparation for demolition, which was undertaken in two phases and completed in November 2014. Many tenants were forcibly evicted from their homes and given compensation at less than 40 percent of the market value through the compulsory purchase processes. In its place, Londoners were promised a shiny new development that would provide affordable, accessible homes for keyworkers desperate to get on the property ladder. The site now known as Elephant Park is being redeveloped by Lendlease into 3,000 brand new homes, a new park which will be Central London's largest for 70 years, retail space and community facilities. In 2002, the council – which sold the land for just £50 million and spent £45 million on preparing for the regeneration – announced that the new site of around 2,530 homes would host 500 social housing units. Yet by the time the successful bidder, Lendlease, unveiled its final plans, *just 82* were put aside for the people they'd turfed out. It has now been claimed that 100 per cent of those sold so far have gone to offshore foreign investors.

In the words of Anna Minton (2017, Kindle Locations 964–971),

> [I]t is worth looking closely at what happened at the Heygate, which was identified as part of an 'opportunity area' for growth in the late 1990s at the same time as it became clear that the homes themselves needed investment. Consequently, the council commissioned a report which concluded that the buildings were structurally sound but in need of complete refurbishment and some partial demolition. It also found that 'the crime statistics show a very low crime rate for this estate.

This estate has a larger than average proportion of elderly people with a significant attachment to the place. The recommendations were for some demolition of the tower blocks and refurbishment of the rest as the most cost-effective solution, environmentally, architecturally and socially. But in 2002, a council document identified the Heygate as 'a barrier to releasing the area's potential', which put in motion the council's plans to demolish it. Using language that a PR company might today have told him to avoid, Southwark's then director of regeneration, Fred Manson, didn't mince his words, arguing that the area needed a different kind of demographic to prosper. 'We need to have a wider range of people living in the borough . . . social housing generates people on low incomes coming in and that generates poor school performances, middle-class people stay away'. He explained that 'above thirty per cent [of the population in need] it becomes pathological' and claimed that 'we're trying to move people from a benefit dependency culture to an enterprise culture', although just getting people to move seemed the main aim as far as the growing number of activists and campaigners were concerned. In 2008, Lendlease were named lead developers. The council's promise of managed but 'inclusive gentrification' soon fell apart as barely any of the promised new homes were built by the time the majority moved out. As the estate emptied and became characterised by boarded-up properties, the Heygate was portrayed by the council as a dangerous and crime-ridden sink estate, justifying the demolition. The media ran with the story, with a BBC piece headlined 'Muggers' "paradise", the Heygate Estate is demolished'. Reflecting a mainstream media narrative which bore no relation to actual crime figures, such commentary has become familiar to estate residents all over London opposing regeneration'.

Source: Based on Minton (2017).

Where policy is designed to benefit others apart from local residents, accountability arrangements tend to reflect this and focus on a wider constituency. For example, in the UK, the model of the Development Corporation was devised to enable control by central government and other interests rather than local residents as it was perceived that the aims were to respond to the needs of the city as a whole, and this would result in a need for a change in population structure.

Governments may want to intervene in neighbourhoods with the aim of helping the local residents and achieving aims of alleviating poverty or inequality. Poor neighbourhood quality may be seen as an unsatisfactory housing situation just as an unhealthy house may be. The discussion above shows the negative impact of a poor-quality neighbourhood on the well-being of residents, and it may be considered that there is a social right to a good neighbourhood as well as a good house.

One problem in designing policy tools to alleviate neighbourhood problems and to help local residents is the importance of the recognition that the neighbourhood may be where the problems manifest themselves, rather than where the causes of the problems reside. The structural issues that influence residential location decisions and the economic and social forces and policies that influence neighbourhood success or decline may have regional, national or global reach. Therefore, the policy response of a focus on the small area may not be appropriate, although this has been the predominant approach. If the source of neighbourhood problems is wider than the neighbourhood, whether due to individual behaviour or socio-economic processes at the national or global level, then it may be that a neighbourhood focus will not be able to tackle the root causes of the perceived problem. An example of this is poverty that may be due to structural forces, such as the uneven impact of global financial markets, that cannot be effectively tackled at the local level. It may be that the concentration of poor people in certain neighbourhoods is the result of the operation of the housing market that stratifies people by their resources. In this case, effective intervention may involve altering the distribution of resources at a societal level rather than focusing on the neighbourhoods that are essentially the symptoms of a wider problem. Despite this drawback we will see that most policy mechanisms are area-based and do focus on the symptoms and not the causes of the perceived problems.

Therefore, area-based interventions have both advantages and disadvantages, although the relative balance of these may change depending on the problem to be addressed and the area concerned. The advantages are the ability to focus on the specific needs of the particular area and to be able to engage local residents and give them accountability for the programmes. Therefore, the area approach may be particularly relevant where the aim is to improve the day-to-day lives of local residents and to respond to their needs for an attractive local environment. However, the disadvantages surround the difficulties in tackling causes rather than the symptoms of problems. Therefore, area projects may find that the causes of the problems experienced by local residents are not amenable to change through a local mechanism.

Policy mechanisms

Because it is the major policy focus, the discussion here is centred on area-based, neighbourhood programmes which are difficult to categorise because there is often a lack of clarity about their purpose. Is the aim to benefit local people or others; to change places or people? This is important, because if the aim is to improve the lives of individual residents, one of the first things that a person who improves their circumstances may do is to move to another area. Alternatively, if the focus is on improving the area, one of the major mechanisms to achieve this may be changing the social balance of the neighbourhood, usually through the attraction of more affluent households. Another source of categorisation could be between physical, social and economic renewal, but many policies consist of a mixture of these elements, although physical renewal tends to predominate, and so we will focus on this here. In addition, we will examine the policies categorised on the basis of the major problems identified earlier. These are segregation and social exclusion.

Physical regeneration

Most regeneration programmes have a physical, social and economic element, and it is sometimes difficult to disentangle these. Nevertheless, it has been common in many countries to focus on change in the physical attributes of a residential neighbourhood, often in the belief that it will bring about social and economic change. Perhaps the first government interventions in housing have been slum clearance schemes designed to demolish areas of housing deemed to be unfit in some way, whether because of the condition of the houses themselves or the perceived neighbourhood features of disease or crime that may spillover to other areas. In such schemes, there may be a concern with the rehousing of the displaced residents either on the redeveloped site or elsewhere, although such a concern has not always been evident, with residents sometimes left to fend for themselves. Therefore, the objectives of physical regeneration schemes may vary considerably and be difficult to uncover and complex.

Criticisms of slum clearance and redevelopment policies have focused on the negative impact on existing residents and the existing social community. This has been reflected in many protest movements against redevelopment. But slum clearance is also seen as expensive and wasteful of existing housing stock that can usually be renewed to appropriate standards through renewal programmes. These can operate at the individual or neighbourhood level and involve the provision of grant aid or loans or direct provision. Case Study 6.2 gives an example of an area renewal programme in Rosario in Argentina that also includes social and economic elements and pursues a twin strategy of helping existing residents whilst encouraging some outward migration.

Case Study 6.2 Rosario Habitat programme, Argentina

With more than 1 million inhabitants, Rosario is an important industrial city and the second largest urban centre in Argentina. Yet more than 100,000 residents live in informal settlements in conditions of extreme poverty, without access to adequate housing or basic services. When the Rosario Habitat programme began in 2000, there were 91 informal settlements in Rosario, covering 10 per cent of the urban surface. Most of these had emerged decades earlier, with rural migration to the main urban centres during the process of industrialisation in Argentina. Others had grown as a result of intra-urban migration.

In addition to physical problems, such as the lack of adequate infrastructure (particularly piped water supply and sewage) and structural deficiencies resulting from homes being built without any technical assistance, the informal settlements in Rosario are characterised by overcrowding, high rates of crime and unemployment and a lack of security of land tenure.

Rosario Habitat is a comprehensive informal settlement upgrading programme carried out by SPV (the local government housing department) in collaboration with a range of partners including residents' associations and other community-based groups, NGOs and national and provincial government agencies. The programme combines infrastructure investments with social development initiatives, and legal actions with measures to create economic opportunities. All programmes are people-centred and aim to build relations among community members. The programme is working in eleven informal settlements in Rosario and comprises the following key components:

- New urban planning of the settlements – including the opening of roads and urban regularisation, ensuring the provision of basic infrastructure (water and electricity supply, sewage, storm drains, gas, paved roads) and community facilities.
- Housing improvements – ensuring satisfactory sanitary conditions through the construction of toilet facilities/sanitary units.
- Building houses with infrastructure – for families living in high-risk areas and/or relocated as a consequence of the new urban planning (relocations should not exceed 30 per cent of the total number of houses in each settlement).
- Legal regularisation – this includes the transfer of state-owned land to SPV, the purchase of private occupied plots, the demarcation of plots and the transfer of legal title to residents.
- Strengthening social networks – with a focus on the direct participation of residents in decision-making processes at all stages (planning, development and implementation).
- Risk prevention programmes – targeted specifically for children and youth (education, nutrition, counselling etc.).
- Employment and income generation – training and work experience for 16–25 year olds and support in establishing social enterprise incubators, primarily benefiting female heads of households.
- Institutional strengthening – training for administrative and technical staff, technical assistance, monitoring and evaluation.

A total of 3,813 families (72 per cent) stayed in the original settlements, with improved housing conditions and infrastructure/services. In a post-occupancy survey, 83 per cent of families stated that they were satisfied with the upgrading process. Beneficiaries especially appreciated improved access to the neighbourhood, new building of roads and passageways, quality of services rendered and increased security.

The remaining 28 per cent of families (1,479) were relocated and have received a house with basic infrastructure and services. Post-occupancy studies show that 86 per cent of families are satisfied with the relocation process. Residents report that their greatest source of satisfaction is having a house of their own.

The programme has generated important impacts in the wider community and city as a whole, with improvements in environmental quality, security, infrastructure,

> public services and traffic circulation. Two sports centres, two community centres, one training centre, three health care centres and three recreational squares have been built and are open for public use.
>
> Source: Based on account from Building and Social Housing Foundation

Neighbourhood renewal policies and programmes have often been pursued as alternatives to slum clearance (for the UK, see Jones and Evans, 2008). In the past, the primary mechanisms have been grants or loans paid to individual owners and landlords in certain designated neighbourhoods that make up a proportion of the cost for certain specified improvements, usually to basic facilities such as bathrooms and structural elements of the house. In the UK, this individual approach was considered to be poor value for money and involved subsidising people to do what they would anyway and so was followed by a more interventionist approach through Housing Action Areas that focused on the neighbourhood as a whole and features such as environmental improvements, as well as structural elements of blocks or terraces of houses as well as individual ones. There was usually an element of new building and renovation through housing associations as well. The final element of this stream of policy was the controversial Housing Market Renewal programme (Allen, 2008). This focused on larger neighbourhoods with 'failing' markets where it was argued that concerted action needed to be taken by the state to stimulate market recovery. However, the programmes resulted in substantial slum clearance activity that was opposed by some local residents and the programme was abolished with the change in government in 2010.

Despite many criticisms of slum clearance, it has proved to be a resilient policy that is rarely absent from a country's policy suite. Slum clearance can appear to offer an easy solution to perceived problems and seems to be popular with some professionals such as architects or housing developers, perhaps because they can benefit from income and profit from the activity. Redevelopment can be a way of changing the social mix in an area through tenure change and can offer possibilities for the involvement of private developers and private finance. Redevelopment may be driven by the wish for higher densities or unlocking the value of the underlying land (see Case Study 6.1). In some cases, it may be driven by the lack of basic infrastructure or even the right to occupy the land by existing residents.

Segregation

As argued earlier, there has been some concern about segregation, whether this is ethnic or social. The response has generally been to try to create 'balanced communities' on the implicit assumption that they are more sustainable than more homogenous neighbourhoods. This assumption is highly contentious, given that most people seem to choose to live in neighbourhoods with people like themselves. Most neighbourhoods are 'imbalanced' and seem to function well, but concern is usually raised when neighbourhoods have a concentration of a disadvantaged group such as an ethnic minority or poorer people.

There are two major mechanisms to achieve balanced communities. The first is physical redevelopment as described earlier, where physical changes are designed to alter the socio-demographic profile, usually by bringing in higher-income residents. The second mechanism is through planning controls. In the UK, there are what are known as Section

106 agreements that require developers to include a negotiated proportion of 'affordable housing' as part of any development. The aim is to 'pepper-pot' below market housing amongst more affluent neighbourhoods to overcome the neighbourhood effects outlined earlier. This result can also be achieved through general planning controls if they are specific enough to allow planning authorities to influence the make-up of housing developments. Evaluations of the Section 106 mechanism have shown mixed results (Atkinson and Kintrea, 2002; Jupp, 1999). Residents from different social classes and in different tenures tended not to mix, mainly because of differences in lifestyle as well as the trend that we tend to make friends with people like ourselves. Nevertheless, the achievement of a degree of social balance tended to reduce social stigma and negative 'neighbourhood effects' for the low-income residents.

Social exclusion

This category is included here because the aim is to change the life chances associated with some neighbourhoods with a concentration of poor people. The aims and structure of the intervention programmes may vary according to the discourse of social exclusion adopted. Emphasis may be on a 'culture of poverty' (Lewis, 1966) where disadvantage is held to be transmitted from generation to generation through the cultivation of minority values that are different from those of the wider society. In this case, interventions are designed to break up the 'mono-cultures' held to be responsible through integration or by changing the community culture and individual behaviour of residents.

Case Study 6.3 describes the Communities First programme in Wales, which was labelled as being designed to alleviate poverty, but was shown by a number of evaluations to lack the policy mechanisms to achieve this. The programme illustrates many of the features and the disadvantages of social exclusion programmes focused on individual neighbourhoods, where the major elements that contribute to the decline of a neighbourhood are located outside the local area and so are not amenable to influence. The programme was abolished in 2017 after successive evaluations showed that it did not achieve its objectives.

Case Study 6.3   Communities First in Wales

The Communities First programme was instituted in 2001 with 142 small areas chosen across Wales with priority given to those in the 100 most deprived areas as measured by the Index of Multiple Deprivation. There was a central support unit in the Welsh Assembly Government and funding given for support units in local authorities and voluntary organisations in the areas concerned. The individual Communities First partnerships were run by a board made up of a "third/third/third" split between the residents, councillors and local and statutory agencies.

Partnerships were encouraged to undertake community capacity building and then to identify needs within their areas and to formulate plans to meet those needs. Objectives of the programme were very general and changed over time but included:

- to build confidence and to raise the self-esteem of people living in the community;
- to increase the incomes of local people (including reducing the costs of food, heat, credit etc.);

- to improve health and well-being;
- to encourage and improve education and skills training for work;
- to create jobs;
- to make communities safe, secure and crime free;
- to ensure public services are delivered in ways which are more responsive and more locally accountable;
- to improve housing and the quality of the environment; and
- to encourage active citizenship.

Communities First was characterised as a capacity-building programme leading to externally funded regeneration and mainstream programme bending; however, it was never made clear how this bending of mainstream funding was to be achieved and the partnerships had no sanctions to exert if the agencies concerned were reluctant to change.

Adamson and Bromiley (2008) point to considerable community development activity in the partnerships with significant levels of community involvement. However, there was frustration at the inability of residents to influence statutory agencies and to achieve programme bending. This means that community capacity building did not result in community empowerment.

All of the evaluations undertaken were unable to identify any significant impact on the relevant outcomes from Communities First. For example, Hincks and Robson (2010) examined a number of indicators in Communities First areas and compared them with other deprived areas not in the programme. Their analysis showed that house prices in the Communities First areas, which were previously lagging those in other deprived areas, had converged with those of other disadvantaged areas not in the programme, but that patterns of change in terms of rates of unemployment and economic inactivity were broadly the same. Between 2001–2008, both Communities First areas and deprived comparator neighbourhoods saw economic activity increase and unemployment decrease by approximately the same amount. The authors concluded that 'the gains that have been made in Communities First areas have been relatively marginal' (ibid. p. 17).

Further, the outcomes varied between types of areas, four of which were identified using a typology developed by the authors:

- *escalator areas* where incomers were of a similar economic status to those already there, but leavers tended to move to more affluent areas;
- *gentrifier areas* where the population changed to a more affluent one as incomers were from less deprived areas and leavers moved to similar or more deprived areas;
- *isolate areas* which had few links with other areas and so movement was small; and
- *transit areas* where most incomers and leavers come from and go to less deprived areas.

Gentrifier areas improved most which seems to indicate that demographic change had been the primary factor in the improvement. Transit areas also improved more than others, further reinforcing the importance of demographic change. Isolate and escalator areas where people move out to more affluent areas were those where little positive change was recorded. In the latter case, this may be that those who benefitted from the programme moved out to be replaced by people from other deprived areas. In this case the area may not improve even though individuals may see some improvement in their lives.

The programme was abolished in 2017.

Source: Based on Clapham (2014).

Evaluating neighbourhood policies

The lack of a clear objective of neighbourhood policies and programmes makes it difficult to evaluate on their own terms. Another complicating factor is that some programmes are clearly not aimed at the well-being of existing residents which is the main criterion used in this book. The example of the Heygate estate in London shows how existing residents can be excluded from the benefits of regeneration, although this is offset by the benefit that accrues to those who are able to afford to purchase the new dwellings and to other council tenants who may benefit from increased maintenance from the surplus generated by the local authority. Therefore, issues of social justice are important here as the groups involved have different income profiles. If the concern is with reducing inequality, as is a major criterion employed in this book, then regeneration programmes such as on the Heygate estate (Case Study 6.1) can be argued to mitigate against this and to reinforce inequality.

However, the example of the Hope VI programme (Case Study 6.4) shows the importance of the context. Here surveys have shown that public housing tenants report increased well-being when they are rehoused with Section 8 vouchers. Also, evidence from the regeneration programme in Rosario in Argentina (Case Study 6.2) shows how a focus on particular low-income neighbourhoods can increase the well-being of residents. However, the example of the Communities First programme in Wales (Case Study 6.3) shows the limitations of area-based approaches in enabling traction on the structural factors that had a major influence on the creation of problematic neighbourhoods. Even if a local regeneration scheme is successful in improving a particular neighbourhood, the positional nature of housing and its location mean that either this is not enough to ensure continuing viability, or that another neighbourhood becomes relatively less desirable and so sinks to the bottom in terms of external and internal esteem.

Case Study 6.4   Hope VI and Choice Neighbourhoods

Hope VI was launched by Congress in 1993 as a federal programme to demolish and redevelop distressed public housing estates. From 1993 to 2010 it funded the demolition of over 150,000 public housing units in 262 projects in 34 states and in so doing changed the face of public housing (Schwartz, 2015). The programme focused on replacing existing estates with low-density mixed developments, often based on the principles of 'new urbanism' and managed locally, and where the new public housing provided tended to be of a higher standard that that it was replacing. Although the overall impact on the image of public housing has been positive, it has resulted in a reduction of the number of units. Only 55 per cent of the units demolished have been replaced by public housing with the remainder being market or less subsidised housing for people who would not qualify for public housing. Not all of the existing residents are rehoused in the new properties with estimates of the proportion varying between 24 and 44 per cent. Others are either rehoused in other public housing estates, find a destination themselves or are given Section 8 vouchers in order to find private rented housing. Surveys of former residents have shown relatively high satisfaction with their new housing (Popkin et al., 2004).

The funding for Hope VI was run down towards 2012 and was replaced by the Obama administration by a new programme, Choice Neighbourhoods, that promised one-for-one replacement of public housing, thus overcoming one of the criticisms of the Hope VI programme.

Note: For a fuller discussion, see Schwartz (2015).

Conclusion

This chapter has described the importance of neighbourhood to residents and explored the changing dynamic of neighbourhoods. The desire to intervene at a neighbourhood level has been an enduring element of government housing policy. From the first examples of slum clearance to recent regeneration programmes, governments have set out to change the nature of neighbourhoods for many reasons. The desire to reduce externalities can be a reason for intervention, as can the desire to ameliorate segregation and social exclusion. Some interventions are designed to change the use of parts of the city and to make profits for developers as well as the aim of privatising public housing.

The examples in the chapter have shown the importance of the context in determining the impact on the well-being of dwellers. What is successful in the USA is not necessarily the same in the UK. But the guidelines in judging the impact of neighbourhood programmes is universally applicable, through assessment of the impact on the well-being of dwellers and the achievement of social justice through the reduction of inequality.

The growing neoliberalism of policy in many contexts has seen the nature of neighbourhood policy change away from intervention to steer the housing market and ameliorate its worst outcomes, towards increasing the privatisation and marketisation of housing. Both the Hope VI programmes in the USA and the regeneration of council estates in the UK are aimed at reducing state involvement in housing, and the well-being of residents is not the primary concern.

Designing neighbourhood programmes to increase the well-being of the existing residents is not easy, because many of the important factors in determining the affordances of neighbourhoods are located outside of the neighbourhoods themselves and so are not amenable to influence. Also, the positional nature of housing means that one group may be helped at the expense of another. These factors mean that intervention at the neighbourhood level can be problematic and reinforces the need to tackle housing problems at source rather than where the symptoms arise.

Chapter 7

Housing supply

Introduction

The aim of this chapter is to examine the processes involved in the supply of new housing. The production of an appropriate quantity and quality of new housing is central to the outcomes of any housing regime. Therefore, housing supply can be an important focus for government housing policy. Three issues have dominated government concerns about housing supply: the quality of housing, the distributional outcomes and the quantity produced. Housing quality issues were covered to a large extent in Chapter 4, and distributional issues will be covered in the following chapter, Chapter 8. In this chapter, the focus is on the quantity of housing and the mechanisms used to influence this. Shortages of supply are likely to lead to problems of availability and affordability and can result in some people being shut out of the market completely.

The chapter starts with an overview of the residential development process and the characteristics of different national housing supply systems. It is argued that the process involves a number of different activities and is undertaken by several actors with particular relationships and objectives – all operating within a context set by government. It was argued in Chapter 4 that governments 'make the market' in housing, and this is also true for housing supply. The particular actors involved, their motivations and relationships are structured by government action in the same way. Some of this structuring may be specific to housing, but other elements may be the consequence of decisions taken in other spheres as housing development is part of the overall economy. Economic policies relating to competition, taxation structures, profitability and so on will impact on the residential development process. As we shall see, the processes of financialisation and the growth of the rentier economy have had important impacts on housing supply processes.

The chapter continues with an examination of the reasons for government involvement in housing supply. It could be argued that governments should leave the market to decide on the amount of housing to provide. However, if there are considered to be problems of housing shortage, and there are questions over the ability of market mechanisms to meet housing need or demand, governments need to be able to judge whether housing output meets the requirements of the population. Therefore, the chapter examines the techniques available to governments to be able to judge whether enough housing exists now and whether future trends will be adequate to achieve requirements in the future. If problems of a shortage of supply are diagnosed, the chapter continues with a discussion of the nature of these problems. Examples are given of a number of neoliberal housing regimes where the nature of the housing supply systems, coupled with policies of financialisation in the economy at large have created systems that are characterised by dominance from a few large companies and problems of shortage and excessive profit taking with an undue emphasis on short-term share value. This is followed by a review of the tools available to governments to intervene in the supply process and an evaluation of the neoliberal demand for a reduction in the regulation in housing regimes.

Supply systems

The process of housebuilding involves the conversion of the inputs of land, labour, materials, finance and technical expertise into the outputs of houses. There are a number of ways of conceptualising this residential development process. One is to elaborate the steps that any housing developer will usually go through in an 'event-sequence model' a very simple one of which is shown below:

1   Project conception and evaluation: involving decisions about the number of houses to be built, costs, availability of finance, feasibility and physical plans
2   Land preparation: if necessary purchase and assembly of a site and any necessary ground preparation works
3   Building construction
4   Marketing and sales: whether the provision of information to prospective tenants and owners and the agreement of terms and conditions of ownership and use

For more complex models, see Carmona et al. (2003) and Parker and Doak (2012).

An important criticism of 'event-sequence models' is that they do not specify who undertakes the different functions or the relationship between them. 'Agency' models focus on the different actors or stakeholders involved in the processes and their relationships (for example, see Fisher and Collins, 1999). These processes may be undertaken by the state (through many different forms of organisation such as local authorities, housing companies, development corporations etc.) or by private companies through market mechanisms or by individuals. In many countries, the development systems are made up of hybrids of these, with the state often setting the framework for other actors in the system. 'Agency' models focus on the resources that these different actors have at their disposal and the way that they interact through negotiation. There may be substantial differences between countries in the relative balance between the three types of agency as well as the precise forms that the institutions involved take. It would seem likely that countries with a neoliberal housing regime would have a development process dominated by private actors, whereas in the social-democratic regime, development would be dominated by government agencies. The corporatist regime would indicate a strong role for third-sector organisations or quangos. Other regimes that have been discussed may also be of interest here. For example, a Southern European regime may be characterised by an individual and family led development process through some form of self-build. Although this could be characterised as a private rather than a public dominated process, it is not a market process. Of course, these categories are ideal types and are over-simplifications that hide many complexities and differences between countries. However, they do open our eyes to the possibilities involved.

There have been a number of models that have attempted to draw together the relevant factors in development (see Healey, 1992). Therefore, residential development has a number of stages and involves the relationships between a number of actors, but it takes place in a particular economic, social, environmental and political context that structures what can and cannot happen. As Parker and Doak (2012) conclude, the outcomes of the development process are socially constructed through the interplay of agency and structure or actors and context. As a result, the differences between countries can be substantial, meaning that each country has its own distinctive supply system with its own features and incentive structures that can have profound impacts on the type, quantity and quality of housing built.

Land is an important element of the housing supply system. It is important to recognise that not all housing production takes place on greenfield sites, but much occurs on sites with other existing uses or involves the bringing back into use of dilapidated or unsuitable existing housing. The supply of land may be influenced by geographical factors such as topography or geological conditions. For example, the housing forms of Hong Kong and Manhattan in New York have clearly been influenced by the availability and nature of the land available, among other factors. The pattern of land ownership may vary between countries as may the rights that owners have over the land. These are set by government as an element of the legal system and governments are, also, usually involved in regulation of the land market. As well as setting the legal and market framework, regulation may take the form of differing forms and extents of influence over the development process through planning controls and/or taxation structures. Therefore, governments in most countries are key players in the supply of land, and Ball (2012) argues that differences in government approaches to the regulation of land are a major factor in differences in the housing supply systems of countries.

The construction process in housing is different from many other goods. The variety and number of elements involved means that the process is often lengthy which increases risk of changing conditions and provides problems in the time-lag between financial investment and the return. In addition, housing production is site-specific because of the locational specificity and fixity of housing. This means that at least part of the construction process has to take place on the final site, thus limiting the place of factory production, although countries may differ in the extent to which factory-based processes have permeated the production process. The disaggregated nature of production also limits the ability to take advantage of technological innovation, although systems may vary in the incentives and possibilities for this.

The nature of the production process, involving different production sites and technologies and labour skills, has led often to sub-contracting being a major feature of many housing production processes. In some systems, there can be a major agent or developer and a hierarchy of sub-contractors beneath this. As a consequence, the management of the production process can be complex and difficult.

Ball (2012) divides the process of housing supply into two phases of development and building construction. This division allows a categorisation of different housing organisational forms. In one form, there is a division between land developers and housebuilders with the developers buying up and preparing land to be sold to builders as individual or larger plots. A second category is residential developers and contractors where the developers contract all or most of the design and building work although they retain ownership of the housing rather than selling it off to builders. The third category is combined developer-builders where firms or other organisations undertake all of the development and construction work themselves, although they may sub-contract actual building tasks. Countries may differ in which of these models is prevalent with, for example, the last category common in the UK and Ireland and the first category more common in the USA and Australia.

Housing is a very heterogeneous product, and many different forms can be produced at any one time and place for different elements of the market. As a consequence, housing organisations may vary considerably in ownership and size. In the UK and other countries such as the USA and Sweden, much new building for owner-occupation is undertaken by national, large-volume housebuilders who undertake the development and building tasks, but there is still room for small companies operating at a local level as well as for government agencies such as housing associations and local councils as well as self-building, which occurs particularly at the high-income end of the market, and for public-private partnerships that occur particularly for regeneration schemes.

The size of housing developers may bring advantages and disadvantages. Large companies will be better able to cope with risk in the development and to raise necessary finance. They may also be able to gain some economies of scale in expertise, purchase of materials and the use of technology as well as be better able to deal with negotiations with government agencies. However, smaller and local developers may be more in tune with local consumer needs and preferences.

All developers have to deal with risk, and this has shaped the nature of the development process. It has been remarked earlier that housing development can take a considerable time and so major investment may have to take place a long time before receipts can be gained. Risks can stem from a number of sources. Economic and financial conditions may change during the development process, for example, interest rates may rise, causing borrowing costs to increase. An economic recession may mean that incomes of prospective buyers may reduce, and so completed properties may not sell or sell at a lower price. Uncertainties also occur about the demands of regulators and the costs of materials and labour that may reduce the financial viability of individual schemes. This can cause problems when the major variable cost that developers face may be the price of land and so developers may pay more for land than is viable when conditions change. Therefore, development is riskier than building, as building costs are relatively stable in the short time that the construction process usually involves. Developers may take actions to reduce risk, such as building a land bank to protect against increases in land prices and undersupplying the market to reduce the risk of unsellable houses at the end of the process.

The above discussion has focused on the supply process in a market situation, and most countries fit this model to a large degree. But there are other development processes such as individual or family self-provision. Self-provision in this form is defined as

> instances where households are involved in the production of their house rather than buying it ready built from a speculative builder or on the second-hand market. This involvement may take the form of building all or part of the dwelling themselves (self-building), or of acting as promoters by bringing together the elements of land, design and construction while not being closely involved in these processes themselves (self-promotion); or of being actively involved in planning and managing the construction process (self-development).
>
> (Clapham et al., 1993, p. 1355)

In countries such as those in Southern Europe a lot of housing outside the major cities is self-produced on family-owned land and rarely changes hands in a market situation. The process can be flexible and geared to the resources of the individual, and it is common for the housing to be completed in stages as resources become available. The process can be helped by the availability of 'off the peg' designs that can be bought and assembled on site or bespoke designs may be made. In many countries of the global south, self-build is the predominant building form as many households construct their housing from the materials and the land that they have available and use their own labour. Governments may aid this process through the provision of finance, materials or expertise. In more developed countries self-build has come to be associated with more affluent segments of the population as a way of getting a bespoke design rather than the standardised products of the major building companies. So, self-provision can be a major supply form and can supplement or replace market provision in different contexts. Some authors (see Ball, 2012) see self-provision as a sign that the private market is being suppressed by state regulation with the implication that it would disappear if the market was freed from state involvement. Others see self-provision as an alternative tenure that maximises individual choice and responsibility and leads to a more personal and higher-quality output. For others, self-provision is just a response to the specific opportunities and constraints faced by households in a specific context.

Also, in many countries, agencies of the state are directly involved in the development process, although the extent and type of involvement may vary considerably. For example, the state may be involved in all or a subset of the development tasks. It may plan provision and commission others, such as private companies, to build it. Or it may undertake the whole process from planning to construction and completion. A state-dominated supply process may be influenced by different factors than the private market one. The role of the profit motive is likely to be less and only restricted to parts of the process. It can be argued that the state copes with risk more effectively than private organisations and is less susceptible to pressures from the cyclical nature of the national economy. Indeed, the production of housing can be used by governments as a counter-cyclical measure to increase employment and economic activity when this is perceived to help the national economy. Investment in housing is an effective way of doing this because it is a relatively labour-intensive process and, therefore, has a direct impact on the incomes and spending power of individuals.

It was argued in Chapter 2 that there are many forms of state organisation, and this is reflected in the many ways that the public production takes place. In some countries, there are national organisations that finance and build housing that are either part of national government or 'arms-length' organisations such as public housing banks. In others, the finance function is undertaken at a national level and the production undertaken at a local or regional level either by local authorities or quasi-public/private organisations such as housing associations. There is little evidence on which of these organisational forms is more efficient or effective, and this is likely to vary according to the context.

Reasons for government intervention

It could be argued that the quantity of houses produced should be left to market processes to determine, but there are many reasons why governments may not accept this view. An obvious one is that supply responds to effective demand, and governments may respond to need as articulated through the political process. But other reasons relate to the uniqueness of housing as a commodity and the functioning of housing markets that may not conform to the neo-classical models of perfect competition. Also, as was argued in Chapter 4, governments make housing markets, and their form and efficiency will vary. If there are many different kinds of supply system, the question arises as to whether there is any evidence that one system is better than others. So how should we evaluate housing supply systems?

Economists would point to supply elasticity as the answer to this question. In other words, how responsive are systems to consumer demand. There is evidence that systems vary widely in this measure. Malpezzi and Maclennan (2001) found that UK elasticities are considerably lower than in the USA (high is good), but that they were much lower, in both countries, post-war compared to the pre-war period (between 1.4 and 4.2 pre-war in the UK and 0.0 to 0.5 post-war compared with 4.4 to 10.4 pre-war and 1.1 to 12.7 post-war in the USA. Caldera and Johansson (2013) categorised OECD countries into three groups of 'highly responsive' (including the USA, Canada, Sweden and Denmark); 'responsive' (New Zealand, Australia, Ireland, Norway and Spain); and 'unresponsive' (including the UK, Netherlands, Switzerland, Austria, Italy, Belgium, France).

Therefore, governments may want to intervene in the housing supply process to correct market failures or to ensure that markets operate more closely to what are seen as market ideals, such as neo-classical conceptions of market functioning. There may be problems of the availability of means of production such as skilled labour or materials or land. The structure of the industry may not conform to ideals of competition and may be dominated by a small number of large companies that erect and enforce barriers to entry to new entrants to the market. The motivations and processes of housing developers and builders may not conform to conceptions of appropriate market behaviour which may lead to what are perceived as unfair profits or levels of remuneration for senior executives. Governments may want to intervene to correct these issues.

Governments may want focus on the outputs of supply systems and how they meet government objectives as well as supply processes. It is worth re-iterating here that intervention is not an unusual activity as governments set the framework within which the housing and land markets operate and which structures the nature of the supply systems and, therefore, their outcomes. Therefore, market failure is equivalent to saying that government has failed to construct and maintain a housing supply system that meets societal objectives. The outcomes of a market that responds to consumer demand may be appropriate if housing is viewed as a market commodity. However, if governments see it as a fundamental human need, then it should respond to need rather than demand, and it is unclear how it can do this if demand and need diverge. In this case governments will want to assess housing need and implement forms of intervention to orient the market towards meeting need. Governments may want to have a mechanism to assess whether the market is operating as they would wish by assessing whether current and future needs are being met.

In many countries such as Argentina or the UK, there is considered to be a housing shortage, and the issue is high on the political agenda. Therefore, there is pressure on governments to intervene to assess supply and to increase it where necessary. There are a number of issues involved here. The first is how to assess whether there is a housing shortage and to identify reasons why any shortage exists. The second is to identify and evaluate the tools and strategies available to governments to increase supply if this is considered necessary.

Assessing housing need and demand

At a national level, assessments of the amount of housing needed are usually based on calculations of existing stock (allowing for a level of empty properties needed for an appropriate level of transactions to take place and the number of unfit dwellings) compared to households and of future requirements based on net immigration, mortality and new household formation. The household formation rate is a key element in this as it reflects the desire and ability of new households to form, primarily by young people leaving the family home to establish an independent household. The rate will be influenced by factors such as the rate of marriage and couple formation and of fertility in terms of the age of the parents when they have children. There is an element of circularity in the household formation rate as the availability of housing is a major factor in the decision to leave home and the ability to establish a new household. Rather as the provision of a new road is likely to lead to an increase in the number of journeys as travel becomes easier, the provision of more housing may increase the propensity of people to establish a new household as it may become easier for them to do so.

Measuring the factors involved in assessing the quantity of housing needed is difficult at the national level because of the difficulty in predicting the relevant variables. In practice, this usually leads to projections based on past trends rather than forecasts based on predictive variables. Barker (2014) criticises the accuracy of population projections in the UK over the past 30 years and notes that birth rates, mortality and migration patterns have often turned out to be marked by considerable uncertainty and the UK is not alone in this regard. Difficulties are most concentrated on household formation rates as argued earlier. Therefore, national assessment of housing need is fraught with uncertainty and difficulty.

At the national scale, the overall level of existing provision and new need may hide imbalances in the type and location of the housing required. For example, differential regional economic growth may result in large stocks of existing housing being in areas where the employment market is weak and shortages may exist in areas of high growth, even if the overall amount is deemed to meet requirements. Change in terms of the location of economic growth or migration or household structure can result in problems as existing housing may be in the wrong place or of the wrong type or size, and so shortages of 'appropriate' housing may exist even where there is an overall surplus.

Another problem is that the factors involved are more difficult to predict accurately at a smaller than national scale. This is because migration between different areas can be difficult to predict and can vary with factors such as the state of the local economy. At the local authority level, institutional boundaries do not always correspond with travel to work areas, and so boundary flows may be significant. The circularity effect may be pronounced here also. If one local authority increases its supply of new housing, it may generate increased border flows from neighbouring authorities. As a consequence of these problems, it has become common in some countries for local authorities to co-operate together in planning for housing supply in order to try to overcome some of the forecasting problems and to share expertise and costs.

A further part of the process of assessing the level of housing needed is to identify the cause of any shortfall that occurs in the quantity of housing produced. It must be remembered that housing is a durable commodity, and so new production in any short time period (such as a year) will be a small proportion of the overall stock. Therefore, increases in the quantity of stock needed may take some time to have a noticeable impact on overall levels. As mentioned in the previous section, housing supply systems vary in their capacity to respond to changes in demand and so a shortage may be due to this.

The techniques above are used to assess the need for housing. However, this does not mean necessarily that any need is transmitted into effective demand. In other words, there may be barriers to entry that mean that households in need cannot access the available housing. The barriers may include institutional rules, but the most obvious barrier in a situation where most housing is provided through market mechanisms is affordability. Therefore, some assessments of the housing required involve an analysis of the price of the new housing compared to the incomes of households. Local authorities may use this information in their regulation of new provision through planning controls. What may seem to be a supply problem may be an affordability problem. Access issues may apply to particular segments of the population or to particular regions more than others.

Also, what appears to be a shortage of housing may be an issue of distribution. For example, in the UK, Dorling (2014) argues that there is now more housing space per person that ever, despite the dominant discourse of housing shortage. He also draws attention to the number of empty and unfit properties and second homes as well as those bought by foreign nationals and never lived in or at least for only part of the year. He argues that the increasing levels of inequality of income and wealth have led to an unequal distribution of housing. Therefore, what appears to be an overall shortage is more a question of the unequal distribution of the existing stock with the resultant problems of affordability and access to suitable housing of specific groups of the population. Where there is a marked difference in income and wealth distribution there may be access problems for parts of the population.

Reasons for low housing supply

There may be a shortage of housing that is due to factors within the housing supply process, and this is the focus of this section. A shortage of supply may be caused by a bottleneck in the supply of resources needed in the production process such as the availability of materials or of appropriately skilled labour. Because in many countries the production of housing is a labour-intensive process that involves the use of a number of different skills, it can be prone to difficulties when these skills are not available. Shortages of finance for developers may exist that prevent them from undertaking the amount of development needed. This constraint may also apply to government agencies that may be limited in the amount of public funding they can commit. The most often cited constraint in many countries is the supply of land for building. This may be because of the land ownership pattern or state involvement through the land use planning system. One of the aims of many planning systems is to prevent development in certain locations and to encourage development in some places and not others according to political views of the appropriate shape of cities and other settlements. Therefore, planning systems will inevitably restrict the total amount of land available, but that does not necessarily mean that enough land is not available in the appropriate locations for development. In some planning systems, the state (usually local governments) has the ability to bring forward land for development and to help developers through land assembly and acquisition. Therefore, the impact of planning systems on land availability is an empirical issue and may vary between systems and countries and different contexts. Stronger planning regulations in the UK, compared with the USA, have been suggested as an explanation for the weaker responsiveness of supply, but there is little evidence that recent reforms to the planning system in the UK have increased supply. For a discussion of the impact of planning controls in the UK, see Bowie (2017) and Barker (2014).

There may be other fundamental reasons why a shortage of housing exists. It was argued in Chapter 4 that housing is to some extent a positional good in that it serves as a status marker. If this is the case, then demand could be almost infinite. As incomes increase, the demand for housing could also increase as households purchase second homes or holiday homes or investment properties as security for older age.

There are other reasons why a perceived shortage may exist in that certain groups are excluded from accessing the housing system or parts of it they aspire to. Attention in some countries has focused on the degree of foreign ownership of housing and its acquisition by investors for capital increase rather than for their own housing or even for rental income. For example, in Inner London a high proportion of new building is bought by investors from abroad, and many do not open the property to the market through occupation or renting out, but keep it empty and hold on to it as an appreciating asset.

Shortages may also be an outcome of the supply system that may result in incentives to developers that result in under-production. It was argued earlier that uncertainty results in developers restricting supply so that they can be sure of selling their output. The long production process of housing means that this is a common phenomenon in many supply systems. Barriers to the entry of new companies to the market may mean that there is little challenge to existing suppliers that could lead to an incentive to increase production. Therefore, long-run shortages can exist as developers seek to ensure that their risk of unsold stock is reduced. Psilander (2012) shows that, in Sweden, over 90 per cent of new private construction is undertaken by the four largest developers, but that small developers are more cost-efficient and can reduce development time substantially. The dominance of a few large developers in many countries such as the UK, USA and Australia has led to housing shortages and a lack of incentives for innovation and efficiency.

Case Study 7.1 is an analysis of changes in the housebuilding industry in major neoliberal regimes, but focusing particularly on the USA. It shows how large companies have coped better with the volatility that characterises neoliberal regimes and have become much more dominant. This has enabled them to erect barriers to entry and to engage in predatory pricing behaviour whilst extracting surplus profits. Financialisation of the economy as a whole has also impacted on the behaviour of these companies leading to an emphasis on short-term appreciation of share value over production considerations. In this situation, dominant companies are in a position to ration the supply of new housing in order to bolster their profitability without facing competitive pressures.

Case Study 7.1 US housebuilding industry

In the years leading up to the collapse of the housing bubble, a wave of mergers and acquisitions hit the homebuilding industry. The resulting concentration of firms remade what had traditionally been an industry of local and regional firms into one marked by companies that strategized at a national, and sometimes international, scale. Although nationally the 10 largest firms held less than a 20 per cent market share, in individual regions their market share was far greater.

Twenty years prior to this, national firms were far less prevalent. As an industry of local and regional firms, few firms were in more than a handful of markets. Between 1989 and 2006, the market share of the largest firms quadrupled, accompanied by an increasing bifurcation between the top firms and others in the industry: whereas in 1989 the top five firms did slightly less than 10 per cent of the total business of the 400 largest firms, by 2006 those five firms did more than 31 per cent, with the top 10 firms responsible for 42 per cent. In local markets, and at the national scale, the industry had changed.

I trace the changes in the homebuilding industry in the 1990s and 2000s, and among the largest homebuilding firms in particular, to make two points. First, housing supply and housing suppliers are not static, nor are the management ideas by which they are governed. I will show that financial transactions and financial firms in the United States became such an important part of the economy in the 1980s and 1990s that managers of the large homebuilding corporations, like managers in many industries, reconfigured their strategies to gain the quarterly approval of securities analysts and shareholders. The goals of growth, production of shareholder value, and corporate streamlining were at the core of homebuilder strategy during this period. Second, that models of housing market behavior, as well as analyses of housing sector growth and change more generally, do not reflect the supply of physical houses and the presence of the homebuilding industry – but they should. Elements such as firm size, market share held by large firms, and other measures of firm dynamics ought to be part of any model.

In a number of countries, homebuilding has become, as an industry, an oligopoly, but "with many opportunities for small firms" (Coiacetto, 2009, p. 33). In the largest English-speaking countries, the industry has a high degree of concentration, not only in Australia and the United States, but even more so in Great Britain, where, Ball notes, there was increasing concentration in the 1990s and 2000s (2012). Ball (2013) also found that the top 10 homebuilding firms in the UK have a much greater market share than their cognate firms in the United States and Australia because of the relative sizes of the territories in question (among other factors). U.S.-based firms not only increased in size and scale, but also adopted a set of supply chain innovations developed at firms like Toyota and Wal-Mart (Knox, 2008). Knox argues that "the increasing dominance of big firms is a result of a combination of factors: access to land and capital, mergers and acquisitions, geographic diversification, improved production methods, product innovation, and strategic alliances" (2008, p. 68).

When linked with the advantages of a dominant market share, greater firm size strengthens the already strong position of the largest companies in an industry. In particular, it enables firms to take on some of the characteristics associated with monopoly or oligopolistic power, including: exclusionary conduct; predatory pricing; raising rivals' costs; predatory buying and competitive discounts; refusals to deal and price squeezes; and investments in competitors territories (Blair and Sokol, 2015; Calvani and Siegfried, 1979). These became common in the build-up to the housing crisis.

How does size benefit firms in the homebuilding industry? Apgar and Baker, in their survey of large builders (2006), found that larger companies had greater access to capital and information; deeper pockets that allowed them to wait out the time involved in navigating increasingly challenging land use regulations; and a more stable economic environment. National homebuilders also have the ability to outbid local and regional developers for choice plots of land (Rybczynski, 2007). Their greater access to capital also has an ancillary effect, in that it enables firms to better weather recessions and thus gain a greater share of the local market, unlike local builders who typically rely on local bank lenders, many of whom can be short of capital when a downturn in the local economy hits (Ambrose and Peek, 2008). When this leads to consolidation in a particular market it can cause, as Coiacetto (2006) points out:

- Above normal prices and profits
- Collusive practice amongst developers
- Less efficient use of resources, especially land
- Exacerbated development cycles
- Less innovation, therefore fewer leaps in productivity, reduced opportunities for sustainable innovations, and less innovative urban environments
- Greater power of developers relative to regulators
- Growing corruption
- The reorientation of the focus and purpose of planning towards more private planning and private sector control over people's lives
- Greater power of developers over the form and structure of cities together with a sidelining of the environmental and public interest in urban development

This can give firms monopoly-like presence in the market. In the 1990s and 2000s, during non-boom times, the 10 largest public builders earned average excess returns of 14 per cent, whereas smaller public firms (all except the top 10) did not (Haughwout et al., 2012). During the boom years (2000–2005), these advantages lessened considerably, with the largest taking in just a 10 per cent premium over all other firms (Haughwout et al., 2012).

Financialization, then, describes our recent economic history, a time during which finance and financial firms became increasingly important within the economy and society more generally. It also led to corporate managers who prioritized decisions that were likely to lead to short-term gains in stock price rather than long-term gain; decreased reinvestment in firms; and increased allocations to stock buybacks and distributions to shareholders, including the managers who were paid in part with stock options. The processes, described as financialization, including the push to achieve increasing shareholder value, drove the rapid growth and consolidation of the homebuilding industry.

Source: Based on Wissoker (2016).

In the neoliberal regime, supply systems are perceived as being part of a 'free market' although, as we have seen earlier, the government is very active in 'making the market' through the establishment and maintenance of the market agents and relationships. However, free markets do not necessarily stay free in the sense that there are incentives for builders and developers to grow larger and for concentration to occur as we have seen in Case Study 7.1. Financialisation has meant that developers have been primarily concerned with shareholder return, leading to an emphasis on merger and takeover activities that boost share price as well as land dealing. In addition, larger companies have shown themselves better in coping with the housing market volatility that is a feature of neoliberal regimes, as they have greater reserves to deal with fluctuations in the market and can spread risk across a number of activities. The result is a low level of new construction as developers seek to reduce risk, without the threat of competition from new entrants to the market. Larger developers can outbid smaller companies for land and have the resources to accumulate large land banks. There is little or no incentive for innovation in materials, technologies or build methods that would make the construction process more efficient. As the British government has pointed out, the result is a 'broken housing market' that produces problematic quantities and qualities of new housing.

Samuel (2018) argues that the volume housebuilders (VHB) in the UK have developed a model of housing production that is driven by the need to generate short-term profit and that this has many ramifications for the standard of design of the housing produced. For example, she argues that

> the form of the public realm in VHB developments is largely dictated by the turning circles of construction machinery not by the needs of communities. The design of the homes themselves are also dictated by the systems that bring them into production and are pared down to the minimum that can be sold – garages are often more symbolic than functional as they are often too small to fit the average car (www.spacetopark.org/). VHBs are reluctant to change their models or to offer more choice as this would incur costs. They are by their very nature conservative and financially risk averse – selling what they know sells, very often in pseudo vernacular styles. Risks associated with environmental damage, dysfunctional communities and health (VHB developments are often built around a car culture on sites with limited public transportation) tend to be passed down the line.
>
> (Samuel, 2018, pp. 4–5)

Tools and strategies of governments

As was outlined in Chapter 4, the state 'makes the market' by setting the framework for the housing supply system (as well as demand) and so is already intervening directly in this system. Therefore, it can be argued that a shortage of supply is a state malfunction. In other words, if the state makes the market, market failure is state failure. This does not imply that supply difficulties can be overcome by removing state intervention as is argued by many neoliberals and by some neo-classical economists. Case Study 7.2 is taken from a UK government Housing White Paper (statement of policy) published in 2017. The case study as a whole shows the neoliberal approach to the housebuilding industry and the perceived supply problems. The ingredients are deregulation and competition in the belief that a 'free market' will deliver more homes. The analysis in this book shows that this is unlikely given the nature of housing. It was argued in Chapter 4, that the nature of housing as a commodity means that a 'perfect market' does not exist and that market imperfections are widespread. Therefore, it is not certain that, left to itself, a housing market would provide the right number and type of houses in the right locations. Indeed, many interventions by government are introduced to overcome supply problems in the market, and it is interesting in the White Paper that the anti-trust element of neoliberalism emerges in recognition of the dominant position of a few large housebuilders and recognition that they create barriers to entry to the market.

Case Study 7.2 UK government White Paper 2017

In February 2017, the British government published a policy statement in the form of a 'White Paper' setting out its view of the housing situation in the country. It must be borne in mind that housing is a devolved function in the UK, and so the policy applies only to England. The White Paper was entitled 'Fixing Our Broken Housing Market' thus giving government recognition to public disquiet about the functioning of the housing market. According to the White Paper, the predominant housing problem was a shortage of housing, and major housing problems, including declining affordability, would be overcome if housing supply were to be increased. We will leave aside for the moment the question of whether there is a housing shortage (it has been argued above that the main problems are of affordability and distribution rather than absolute shortage), and here focus on the government view of why there is a housing supply problem and their view of how it can be rectified.

The White Paper divides the factors behind the housing supply problems into three categories. The first is the perceived shortcomings of the planning system in not designating enough land for housing. Therefore, there are proposals to force local authorities to make more land available and to simplify and speed up the planning process. There are also proposals to avoid arguments about the extent of new housing need in a locality through the use of a standardised method for assessing housing need. In addition, there are proposals to limit the ability of local planning authorities to reject development on the grounds of high density and to rationalise building and space standards for new housing. In essence, this category blames land use planning and other forms of regulation for supply problems and puts forward proposals for deregulation and simplification.

The second category is concerned with speeding up development and again the major focus is on the land use planning system with planning departments being urged to make decisions more quickly. Developers are helped by reducing the scope for planning conditions on new developments, simplifying regulations about the protection of wildlife sensitive sites and reforming the payments developers make to development costs. Also, government promises to help overcome skills shortages in the construction industry by reforming its training provision whilst urging building companies to invest more heavily in training.

The final category is what is labelled 'diversifying the market' in which supply problems are put down to the structure of the house building industry. The problem is perceived to be the domination by ten large companies that is said to hinder competition, give little incentive to innovation and reduce consumer choice. The White Paper states (p. 46):

> There is a lack of competition. We increasingly depend on the major house builders to build most housing. Smaller firms bore the brunt of the recession and their output still falls far short of pre-recession levels. . . . The business model for many commercial developers limits the number of homes that are built. The 2007 recession reinforced cautious behaviours at all stages of the house-building process. Major builders rely on sub-contracting, which pushes innovation and risk down the supply chain to those least equipped to respond. House-building methods have barely changed in over 100 years.

> This view contradicts the Office of Fair Trading (OFT, 2008) which concluded that there was little evidence to suggest that the market for newly constructed homes was not competitive and pointed out that, although the industry is dominated by a small number of large firms, the industry is characterised by takeovers and mergers.
>
> Therefore, this category resorts to the anti-trust element of neoliberalism that has been evident in competition policy in a number of fields. The case study as a whole shows the neoliberal approach to the housebuilding industry and the perceived supply problems. The ingredients are deregulation and competition in the belief that a 'free market' will deliver more homes. The analysis in this book shows that this is unlikely given the nature of housing. It is interesting that there is little mention of the land market in the White Paper and of the importance of house price volatility in restricting supply. Nor is there any recognition that it will take a substantial increase in new supply over a large number of years to impact on price as it is a small proportion of existing supply. Distributional issues are also not considered.
>
> The major shortfall in the historical level of new housing construction has been caused by the lack of new council housing. There is little help for this to be resumed. Housing Associations, before 2010 seen as the major provider of social housing, are now seen as developers for private market rent with little or no public subsidy.

In countries with a neoliberal welfare regime it is likely that there will be strong political calls for the removal of existing state regulation in order to 'free up the market'. The kinds of mechanisms involved are usually the regulations on housing quality discussed in Chapter 5 and requirements put on developers through the land use planning system such as the need for planning permission or the payment of a contribution towards infrastructure developments. Any price controls such as rent controls may also fall into this category as they reduce the amount that developers or landlords can receive for their investment. However, reductions in income may be recouped through lower land prices for developers and may not necessarily reduce the amount of development that takes place.

One of the major ways in which governments are involved in the supply process in many countries is through a land use planning system to regulate the amount, type, form and location of housing development. The importance of government intervention in the supply of housing through land use planning is worth some consideration as it is often a controversial element of the process in many countries and is often cited by developers and some politicians as a major reason for deficiencies in supply. The justifications of land use planning relate to the specific nature of housing as a good as described in Chapter 4. It was argued earlier that the costs and benefits of housing development may be different for the individual developer or consumer to the society. Land use planning is in many countries a major way in which governments (both local and national) attempt to impose the perceived public interest over that of the individuals concerned. Therefore, planning controls can be perceived as an interference in the workings of the market, and in many neoliberal housing regimes, they have been minimised, whereas in social-democratic regimes they have been seen as a crucial mechanism for regulating housing markets to ensure that they meet social objectives.

One of the major considerations in planning decisions is the impact of housing on the shape of cities and other settlements. A major objective for the establishment of the British town planning system was the perceived need to avoid urban sprawl and to contain the spread of towns and cities. The conservation of the countryside and of land important to wildlife, landscape or lifestyle amenity was also an important objective. Housing development can have many externalities both to governments, who may be responsible for infrastructure development and the provision of public services that may be needed by residents such as schools, and to other individuals and communities who may see their amenity reduced by a development. The term 'nimbyism' has been coined to refer to the efforts of some people to object to some developments because they reduce their amenity. The planning system then becomes a forum for the reconciliation of the different interests involved in development. The land use planning system may also be used by governments to achieve societal interests that may not be in the interests of individual developers. Perhaps the most important area where this applies is in the climate change or sustainability agenda as we will discuss in Chapter 11, but there are many other examples. Therefore, the land use planning system is a major tool for governments in achieving their aims in the housing system and will shape the supply process to varying degrees. Of course, this intervention may lead to different outcomes than markets would lead to if individual interests were allowed to reign as it is designed to do. However, the planning process may also lead to unintended and undesired outcomes such as delay in development, increased risk for developers and to less housing being produced than desired.

Regulation by government may also take the form of influence on the design of the house or the materials used or elements of the structural integrity as discussed in Chapter 4. Here, governments may be acting in the interests of consumers and others in ensuring that harm is minimised to residents and others through market externalities.

In general, the removal of government regulation in an attempt to increase the supply of new housing may involve the sacrifice of other objectives. For example, a common government response to a housing shortage has been to reduce quality standards resulting in smaller and lower-quality housing being produced. As Samuel (2018) points out the standard of housing produced by the volume housebuilders in the UK already has very low design standards. The reduction of planning controls may result in a lack of influence over the shape of cities and involve sacrifice of sustainability objectives which the planning system is crucial in achieving (see Chapter 11). It may be that a judgement is made that these sacrifices are worthwhile, but there are further questions about the utility of deregulation in increasing supply.

For example, there is an interesting debate about the incidence of these government regulatory interventions that mirrors the debate on the incidence of subsidy. If government decides to reduce the costs that developers face by reducing regulation, then this increases the income that developers receive. They may hold on to this in the form of increased profits, or they may be in a position to pay more for their land in competition with other developers. The latter action would not necessarily result in more output as the benefit would just add to land costs. In effect, the beneficiaries would be landowners. The same argument is often made over subsidy where any increase in funding say through demand-side subsidies would just be absorbed through higher land costs (see Chapter 8).

A similar argument can be made if the problem is perceived to be caused by a lack of demand. In this case, intervention may be designed to increase demand and may take the form of direct lending or action through banks and other lenders, either by creating new financing mechanisms, or the provision of subsidy. However, it is unlikely that this form of intervention will have the required effect in more than the short term because of the impact on house prices and then on land prices.

So rather than deregulation or demand subsidies, change may be more effective if it is directed towards altering the nature of the supply system. This could take the form of changing incentives to developers through taxation or subsidy systems or company regulation to change the make-up or behaviour of developers. An example of this may be underwriting risk as it was argued earlier that the development process is inherently risky and takes a long time. Governments are in a good position when compared with individual developers to be able to cope with risk as they can more easily absorb any difficulties or costs that may arise because of their large scale and ability to raise revenues through taxation. Governments could act to reduce uncertainty and to reduce barriers to entry where they exist.

Government may also have an important role in helping to overcome any bottlenecks in resources. For example, governments and government agencies are usually owners of large quantities of land that could be made available for housebuilding. Planning powers could be used to ensure that more development land is available, and sites could be made more amenable for development through the removal of previous uses or contamination or through the provision of infrastructure. Governments also often have powers of acquisition that can be used to aid site assembly that developers may find more difficult. Labour shortages can be overcome through training and other labour market policies.

One policy tool available to governments is to directly produce housing themselves and so supplement the efforts of private developers. For example, in the UK the private provision of new housing has been fairly constant over time, but the variable has been the production of public housing that is presently at an all-time low. The private sector output has not expanded enough to make up the shortfall. It is interesting that China has chosen to build large amounts of public rented housing in the major cities, not from a political ideology, but in order to solve the shortage and affordability problems of migrants to the city – a problem that was hurting the economy.

Conclusion

Each country has an individual housing supply system that varies in terms of the agents involved and their relationships and motivations which are major elements in the housing regime. As a result, housing regimes vary in the amount and type of housing output and the efficiency and responsiveness to demand of the supply system, as shown by the different elasticities outlined in the chapter. However, there are some general trends in neoliberal regimes that warrant comment. There has been substantial concentration of market share in the hands of a few large companies in many neoliberal housing regimes with the consequent erection of barriers to entry and reduction in incentives to increase efficiency or output or design quality. Financialisation has resulted in a focus on short-term share value and the monopoly position has allowed surplus profits to be made.

The blame for supply problems is usually laid by neoliberals at the door of existing government involvement through the land use planning system. There may be ways that this process in many countries can be improved to make the development process more efficient and increase supply, whilst achieving the societal objectives that the planning system seeks to pursue. But it is clear that this is not the only answer to supply problems and that the state can intervene to deal with the defects in the housing market.

A number of different forms of intervention have been described in the chapter, and their usefulness will depend on the characteristics of the particular supply system. The methods include regulating the market to ensure competition, investing in training and other manpower policies to offset labour and skills shortages, as well as investing in infrastructure provision to enable new development. Overcoming constraints such as bottlenecks in the supply of land or materials or labour skills can be the focus of government action, as can the provision of public land and infrastructure. Action to underwrite risk and uncertainty may generate developer confidence. Supply-side subsidies may overcome cost constraints on supply (see Chapter 8).

A further method of intervention is the direct provision of housing by the state. In the UK, the decline in new house building has been because of the substantial reduction of council house building as the private sector has not increased production in compensation. China is an example of a country where housing shortages in the major cities are being met through direct state provision, and many other countries could follow this lead in order to reduce volatility, increase the quality and quantity of supply and ensure affordability (see Chapter 9).

Chapter 8

The distribution and affordability of housing

The aims of this chapter are to review the evidence on the distribution of housing between different social, economic and demographic groups and the factors that impact on this, as well as the strategies that governments can use to influence this. Many governments intervene in housing markets in order to change the distribution of benefits in the pursuit of a concept of social justice. Housing not only reflects the existing distribution of wealth and income, but serves as a mechanism to generate and reinforce existing inequalities because of the differences in housing costs and the impact of rising house prices.

The chapter starts with a review of the factors that influence the distribution of housing such as the overall distribution of income and wealth in a country. The way that these wider factors are translated into housing outcomes is examined by using the concept of 'affordability' to assess the kind of housing that different groups are able to access. The chapter then examines the factors that underlie government intervention in this area such as concepts of social justice and 'fairness' and 'equality' and their meaning and impact in the housing sphere. Government attempts to alter this distribution through subsidies and taxation are described and evaluated and conclusions drawn on the impact of current neoliberal trends.

The distribution of income and wealth

Where housing is accessed primarily through market mechanisms, the total amount and the distribution of income and wealth in the society is bound to have a profound influence on the distribution of housing. Also, because housing is such an important element of personal wealth and one of the biggest items of household expenditure, the distribution of housing also influences the distribution of income and wealth.

Governments are usually concerned with the overall amount of income in the society as measured by the gross domestic product (GDP), and this will influence the total amount of income that can be devoted to housing (see Table 8.1). Countries also vary in the proportion of national income that is spent on housing which may impact positively on housing standards but may have a negative effect on the prosperity of other parts of the economy. Housing may also play a positive part in influencing the national income as it may provide the infrastructure to enable the labour market to operate well to support economic activity.

However, here we are primarily interested in the distribution of the overall national income between individuals as this will influence the access that different people will have to elements of the housing market and their purchasing power within it. There are many ways of measuring the distribution of income and of making comparisons between countries.

Table 8.1 Gini coefficients for six case study countries, 2015

	Gini coefficient	GDP per head $ 2016
China	0.465	6,894
Argentina	0.426 (for 2014)	20,451
USA	0.39	57,591
UK	0.36	42,651
Australia	0.337	47,769
Sweden	0.274	49,075

Source: OECD, World Data Bank, Statista

The most widely used measure is the Gini coefficient. However, as Atkinson (2015) points out this provides a single-figure, summary measure of inequality of incomes that involves weighting between people in different places in the distribution and so is inherently subjective in the sense that the use of different distributional weightings would give a different figure. Nevertheless, the Gini coefficient is widely used in comparisons between countries and over time.

Table 8.1 gives the coefficients for the six countries used as examples in this book, and it is evident that there is a wide variation. When examined over time, in the UK and the USA there was a trend of a more egalitarian distribution, as shown by a falling coefficient, from the 1940s to the 1980s, but then a growing disparity of incomes that has been called the 'inequality turn' from the 1980s to the 1990s, although it has been relatively stable since. Because of the rather arbitrary nature of the Gini coefficient, the other widely used method of measuring inequality is to compare the share of different parts of the distribution. For example, in the UK the share of the top 1 per cent of earners was 19 per cent in 1919, falling to 6 per cent in 1979 and since has more than doubled. Brown (2017, p. 5) concludes that,

> What we are talking about in fact is a marked and increasing imbalance between the contributions that various groups and individuals are making to society, on the one hand, and the economic and other benefits they derive from it, on the other.

The changing pattern reflects the changing nature of the labour market as well as the differences in returns from labour and capital (Piketty, 2014). The general picture in the UK and other globalised countries is a distention of the labour market with a large increase in incomes of the highest section with a growing lower end and a middle that blurs the difference between skilled manual and middle-class jobs.

The distribution of wealth has also become more unequal in many countries (Brown, 2017). Piketty (2014) shows how a number of mechanisms, including the increased return to capital rather than labour have meant that the top 1 per cent have increased their share of total capital (for a review of the impact of Piketty's analysis on housing, see Maclennan and Miao, 2017). This phenomenon has been labelled as 'rentier capitalism'. Standing (2017, pp. 4–5) defines a rentier as 'someone who gains income from possession of assets rather than from labour'. A rentier state 'has institutions and policies that favour the interests of rentiers' and a rentier economy is 'one that receives a large share of income in the form of rent'.

The precise changes in income and wealth vary between countries depending on their unique economies, demographics and history. But, does inequality matter? There are two types of answers to this question. The first type relates to political ideologies and the impact on the wider society which we will cover in the next section. The second type considers the impact of inequality on specific factors such as our interest here in the housing situation, and we will focus on this in a later section.

Concepts of social justice

There are a variety of views on the desirability of different distributions of income and wealth that are embedded in political philosophies and set the context for government intervention in general and in housing in particular. A common distinction is between a concern with either equality of opportunity or equality of outcome. Equality of opportunity focuses on evening out the opportunities that individuals have to make their way in life and tends to accept that different abilities will result in different actual outcomes. These differences in outcome are held to be a key motivational element for individual effort and achievement. The major problem with a focus on equality of opportunity is that it is impossible to sustain in practice. To take a hypothetical example, the first participants in the housing market may be equal in their opportunities, but once they achieve unequal outcomes these will be transmitted to future generations through gifts and inheritances and the succeeding entrants will not start out equal. Countries with most unequal distributions of income and wealth tend to have less social mobility (OECD, 2008). As Deaton (2013, p. 107 quoted in Brown, 2017) argues: 'Even if we believe that equality of opportunity is what we want, and we don't care about inequality of outcomes, the two tend to go together, which suggests that inequality itself is a barrier to equal opportunity'. Atkinson (2015) points out that a focus entirely on equality of opportunity, which is politically popular and the general neoliberal position, ignores the fact that outcomes matter as well, both for the life chances of households today and over time as inequalities of outcome will result in unequal opportunities in future generations.

Therefore, equality of outcome is necessary to achieve equality of opportunity in the long term, because it is possible to transmit benefits between generations as happens in all countries through the transmission of housing wealth. The position is further worsened when the gains to capital through increases in house prices outrun increases in earnings. Disadvantage can then become entrenched and cumulative. This situation is further complicated in housing because of its nature as, at least in part a positional good (Hirsch, 1977). For example, the first owner-occupiers in say the 1930s were able to enjoy the status of a small elite and the use of a house that brought the affordances of suburban living with low densities and green space nearby. Following owner-occupiers do not get these same advantages as they are part of a larger number, and so the status is less, and the advantages of the suburban location have been devalued through a degraded popular use. The first owners see the affordances of their ownership also degrading as it becomes more popular, and so it is not surprising that they attempt to protect their position through nimbyism and political choices to restrict housing output and sustain house price increases. The newer entrants to owner-occupation can never gain the benefits that the first entrants did, because the nature of the good has been changed by its popularity. This analysis is reinforced when house prices increase faster than incomes, so that succeeding cohorts find it more difficult to access the tenure and more expensive if they manage to do this. Concerns about generational divides in housing consumption need to be seen through this lens as they are rooted in the maturation of owner-occupation and its positional status.

A concern with equality of outcomes will also accept the importance of equality of opportunity but will also focus on the eventual outcomes. This concern can be divided into two categories. The first is a focus on the situation of the poorest. In this category, the issue is poverty and the incomes of those on the poverty line, however this is defined, and whether it is absolute or relative. Poverty can be defined in terms of a state benefit level or an absolute 'basket of goods', i.e. the amount required to sustain a given level of consumption of goods deemed to be necessary, or relative to the incomes of others. In this last method, the poverty level can be set as a given proportion of the mean or median income.

The second major category is a focus on the whole of the income spectrum in a concern with inequality in general. Life chances are dependent both on the absolute level of income in a society and also the relative level. The absolute level will have an impact on the possibility of the lowest-income groups accessing the necessities of life, amongst which is adequate housing. It can be argued that the greater the distance between rich and poor the more likely there is to be poverty. But there is also a strong argument for a focus on inequality and relative poverty. Many goods are positional goods in that their importance rests in the identity and status and the relative consumption benefits that they give to consumers. As we have argued throughout this book, housing is one example of a positional good, in that relative consumption is important in public display of low status leading to identity problems and an inability to lead a 'normal' lifestyle as well as mediating access to a number of private and public facilities including employment, education and health care and so on.

Atkinson (2015) argues that high levels of inequality (although this is not defined) can lead to a lack of social cohesion in a society, and he cites the work of Wilkinson and Pickett (2010) who show that inequality leads to increased social problems such as crime, obesity, poor health, teenage pregnancy and many other problems that have a social gradient within societies, that is they are more prevalent among people at the lower ends of the income scale. They argue that inequality 'gets under the skin' in influencing self-esteem and identity and so leads to the presence of the most problematic social element of shame which has large consequences on mental health and general social functioning. Atkinson (2015) also quotes the IMF who argue that inequality reduces economic growth because of the disincentive effects on lower-income people. Brown (2017) quotes a number of IMF and OECD analyses that show that there is an inverse relationship between the income share accruing to the top 20 per cent and economic growth. Lower inequality is strongly related to faster and more durable economic growth and redistributive efforts from governments have little negative impacts on growth. This is despite the justification often used for inequality that it provides the incentives to high-income individuals that are the drivers of growth. Rather, the extent of inequality has been a major contributor to the 'secular stagnation' that has engulfed the UK and many other countries since the 2008 Global Financial Crisis.

Brown (2017) outlines a number of reasons for the increase in inequality in many (particularly neoliberal) countries. He divides them into two categories of structural causes that are usually wider than individual countries and institutional factors that are dependent on the political and societal choices made by governments. The structural causes include: globalisation, skill-based technological change, changes in the organisation of work, changes in household structures, winner-take-all markets and the inherent nature of capitalism. The institutional factors include: neoliberal government policies, the changing role of labour market institutions, financialisation and rent seeking and government macro-economic policies. There is not the space here to discuss each of these in the depth they deserve. There are two general points we can draw for our present task. The first is the general structural context that surrounds housing policies and the constraints this puts on the capacity of individual governments to meet their own objectives. However, individual governments do have some scope for choice, and it is clear that neoliberal governments have chosen to adopt and to reinforce the wider changes. The differences in the choices made and the reactions to wider forces are likely to lead to different housing regimes that will vary according to the definitions of social justice adopted by governments and the importance they give to them.

Housing and income inequality

A household's income will influence their ability to access housing. Obviously, their wealth is also important, and this will be considered in the next section. The income will be a product of savings, access to funds from other parties (such as parents or other family members), subsidies and income from employment to pay rent or repay a mortgage. As housing is such an expensive good when compared with income, most people need to access borrowing through a mortgage, at least at some time in their housing pathway. Therefore, the policies and practices of lending institutions can play a key role in influencing access. As outlined in Chapter 4, governments make the rules under which lending institutions operate, and this regulation is a key element in the armoury used by some governments to influence outcomes in the housing market.

A common concept in the analysis of housing outcomes is that of affordability which can be defined in a number of ways (see Leishman and Rowley, 2012, for a review). The concept of affordability is here taken to be the ability of a certain household to be able to meet their housing costs, but the concept has been used in many ways. It is often used in relation to the difficulties of access that certain groups may have to housing of their choice. Commonly, this has meant the difficulties of access to owner-occupation. Also, the term 'affordable' is often used in relation to particular forms of housing that are below market price such as subsidised public or social housing. In the UK, it is specifically used to denote housing provided by private developers at below market prices as part of their planning obligations for new housing developments. However, our definition here is more specific and is related particularly to the ability to pay for housing.

Using our concept here, perhaps the most common national-level indicator of affordability is to compare average house prices with average earnings. This method can be useful in charting changes over time and highlighting differences between countries, but the focus on averages could hide disparities between different income and other groups in the society. To overcome this problem, Meen et al. (2005) have suggested that the focus should be on the comparison between lower-quartile owner-occupied house prices and lower-quartile household earnings as an indicator that is more useful in showing affordability problems. However, the focus on house prices ignore tenures other than owner-occupation and is not an accurate predictor of housing costs as these may be influenced by other factors such as the level of interest rates. Therefore, a more accurate yardstick is to measure housing outgoings as a proportion of income. Some countries use a figure such as 30 per cent as a maximum for housing to be affordable to a household, and this can be the threshold for subsidy payments. The problem with this measure is that a given percentage is arbitrary and does differ between countries. Also, it fails to take account of differences in income. Thirty per cent of a low income can be problematic to live on, but the same proportion of a high income may still leave enough for a lavish lifestyle.

An analysis of affordability can be made on a national level, but it is also common to make comparisons for a specific location such as a city or region. Average house prices are compared with average incomes to give an indicator of affordability. The comparison can be varied so that, for example, the lowest decile of incomes is compared with the lowest deciles of housing costs. This can give a clearer view of the possibility of different income groups being able to afford housing.

However it is done, the comparison of incomes to housing costs is essentially arbitrary because there is no shared understanding of what percentage of income is affordable. It will vary between people of different income levels, as mentioned earlier, but also depends on the length of time a high level is sustained. For example, a young professional couple at the start of their careers may decide a high level is sustainable for a few years on the understanding that their incomes will increase and housing costs will decline over time in real terms with inflation.

Some of these problems can be overcome by using the residual income approach. Here housing is said to be unaffordable if the income after costs is lower than the set poverty level. This measure allows for the differences in income, but its value is dependent on the utility of the poverty level and how this is defined and measured. If the poverty level is not enough for people to live on, then the residual is not sufficient and so would need to be revised. Therefore, the measure is still essentially arbitrary and will vary between countries unless the residual is compared to the cost of a defined 'basket of goods' that could be constituted similarly across different countries and may overcome problems of different poverty lines. But even here there needs to be universal agreement on the appropriate 'basket of goods' that would cope with differences in consumption patterns between countries. It will also vary between the quality of housing chosen which will be reflected in its cost.

A further issue is the differences in household composition that could impact on the need for housing space and the household income. The point is that households are not the same in their need for space and their capacity to earn income even if they are of the same size.

All definitions of affordability outlined above face two intractable problems. The first is that they pay no attention to the quality of housing being paid for. A dwelling may be affordable, but how does it compare to minimum societal requirements or more general quality judgements? So, change in housing affordability may mask changes in housing quality. The second problem is the lack of a focus on the strategies employed by households in relation to their housing costs and their housing pathway in general. Households may choose a low housing quality and cost in order to maximise expenditure on other requirements such as food or clothing. Others may choose to pay higher housing costs because it provides access to good schools or provides a better context for family life. To avoid these problems one is left with a complex definition of affordability as put forward by Stone (2006, p. 151): 'an expression of the social and material experiences of people, constituted as households, in relation to their individual housing situations'. However, this definition is difficult to operationalise and takes away a focus on housing costs.

Affordability may be defined in different ways in political discourse. For example, in the UK and in many other countries, there is said currently to be an affordability problem for young people. What is meant by this though is the difficulty that many first-time buyers are having in accessing the tenure, even if they are adequately housed in the private rented sector at a rent that is equivalent to what they would pay to service a mortgage. In this case, it is more accurate to label the problem as one of access rather than affordability. The problem can only really be defined as an affordability issue if the costs are related to what people aspire to, or what others think they deserve, or to what they believe is fair. In many countries, such as the UK, there are cultural and financial biases towards owner-occupation, and many young people want to follow their parents who they have seen benefit from this situation. So, the problem of affordability in political discourse may not be necessarily focused on the poorest sections of the population, but may relate to any group that are perceived to have a problem accessing the housing that they and the government see as deserving. This means that government attempts to deal with problems of affordability may be focused on particular groups, particularly those on whom they may rely for political support.

Affordability issues may have a profound impact on the lives of individual households. At the extreme they may result in the loss of a dwelling and homelessness if households are unable to keep up their rent or mortgage payments. Another result could be the lack of other vital necessities such as food or clothing if housing costs take up a large proportion of disposable income. In addition, affordability problems may impact on a housing pathway at a later stage. For example, a young couple unable to afford their mortgage payments may have to remain in renting for their whole housing pathway and may not accrue wealth to fund their care in old age. Difficulty in meeting housing costs may have an impact on a person's enjoyment of their house. One of the constituents of a home is the feeling of security and lack of worry, and this is unlikely to obtain when there are concerns about the difficulties in paying housing costs and the implications of any default in payment.

The income of a household (whether an individual or a number of people such as a family or other group) is a major factor in their ability to afford housing and the quality and type of housing that they can access. In the owner-occupied sector the policies and practices of banks and building societies are also important in translating income into the capital required to purchase a house. These may vary over time and reflect government financial regulation as well as country-specific institutional factors and views of the current and future state of the economy. General interest rates will also impact on the cost of a loan or mortgage and the different forms of these were covered in Chapter 4. Governments are crucial in structuring the financial system for housing, as was discussed in Chapter 4, but they also may want to intervene to improve affordability for all or for particular social or economic groups. The provision of subsidy is one way that they can do this, and we will turn to this later.

When purchasing a dwelling, the wealth of a household may also be important, and it is to this that we now turn.

Housing and wealth

In many countries wealth is distributed more unequally than incomes (for a review, see Brown, 2017). There is a reciprocal relationship between housing and wealth. Wealth can provide access to housing, with most first-time buyer households in the UK now receiving gifts from parents that enable them to access owner-occupation. This of course provides a mechanism by which existing inequalities in wealth are reproduced, as households without this family wealth may not be able to access the tenure or to buy more expensive property and so build up wealth. Therefore, wealth of the individual or family is both an important factor in influencing housing outcomes, but housing itself is also an important constituent of wealth. It is often the personal wealth accrued through housing that is used to provide a stepping stone to that wealth for a subsequent generation and so reinforcing inequalities, especially if the greater returns to wealth than to labour identified by Piketty (2014) are taken into account. The effect of this is to create intergenerational inequalities that are at the heart of many affordability concerns today. If the returns on wealth generated through housing are greater than those from employment, then older people are going to be increasingly sitting on an asset that their children are not going to be able to afford to purchase without their help. Then if those older people are wanting to hold on to that asset, partly to enjoy the utility of their possible longstanding home, but also to help fund their own old age through equity release as public care and health services are removed through 'austerity', then the transfer of wealth across the generations is not going to be simple. In these circumstances, wealth may not be transferred but consumed, or it may skip a generation as people are living longer. As a result, many young people may not be able to enjoy the access to owner-occupation and the wealth accumulation that their parents did. Therefore, intergenerational inequity is a result of the playing out of the neoliberal housing ideology rather than the greed of an older generation as it is sometimes presented in the media and political debate.

Inequalities in wealth may have an important influence on housing outcomes as well as the reproduction of inequalities across generations. Because of its high cost, the ownership of a house is the greatest constituent of personal wealth for many people and in many societies. Therefore, governments may want to influence this. The accrual of housing wealth raises important ethical and political issues that often conflict. It is where basic principles of the rights of property ownership clash with concerns about fairness and inequality. For example, many politicians will argue that a family has the right to look after its own and every parent the right to help their children by passing on an inheritance. But of course, doing this exacerbates inequalities of both opportunity and outcome (Maxton and Randers, 2016). Housing as a major constituent of personal wealth is at the crux of this issue. In addition, households may want to accumulate wealth in their houses to see them through times of lower income such as old age and many governments have encouraged this process through ideas of asset-based welfare. But again, this will promote inequalities especially as many households will either not have access to this as they are renters, or own assets that do not appreciate at the same rate as others or at all. Therefore, it is no surprise that questions of housing wealth are hotly disputed and are at the heart of different political ideologies and different housing regimes.

Government intervention

Affordability problems may receive a lot of attention from the media and may impact on groups that governments and other political parties may depend on for political support. The example of intergenerational equity was mentioned earlier as one particular problem for governments as many young people have not had the access to owner-occupation that their parents did. In addition, governments may be persuaded to intervene on the basis of political philosophies that contain views on desirable levels of inequality in income and wealth and its reproduction between generations. As a consequence, affordability issues may loom large in the perception of many governments and intervention be considered important.

Interventions can be split between those that help functioning of the housing market, that is those aimed at 'making the market' and overcoming some of the problems identified in Chapter 4 and those aimed at achieving distributional goals. The latter may be divided into those aimed at housing outcomes and those aimed at wider distributional issues and at the wider economy, such as fostering economic growth or acting as counter-cyclical mechanisms.

The rationale for government intervention in affordability may differ. One element may be electoral calculation if support is required from a particular affected group. But the reasons may also relate to political philosophies and their inherent concern about issues such as inequality and concepts such as fairness. In a market housing system, it is likely that any inequalities of income and wealth will be reflected in housing outcomes and that, because of its high cost and its increase in exchange value it will be a major mechanism for the reproduction, sustenance and perhaps exacerbation of existing inequalities. Of course, any government concerned about the extent of inequality could intervene more directly through general taxation and benefits policies as well as employment and economic policies to reduce inequality in incomes. It could be expected that such interventions would have an impact on the capacity of households to afford housing and so obviate the need for direct housing intervention. However, governments may perceive problems with the direct approach, perhaps because of its impact on work incentives or on beliefs about independence or laissez-faire in the economic field and so tackle inequalities in housing more specifically by focusing on the housing system itself. 'Social welfare arguments represent a key justification for state intervention designed to promote a defined minimum provision of housing at a defined affordable price' (Leishman and Rowley, 2012, p. 394).

So, what impact does inequality in housing have? Clearly it could have a massive impact on those on low incomes and in poverty. But inequality of housing outcomes may have an impact on social solidarity. Differences in housing may provide a very physical indicator of inequalities that are more evident than differences in income. We rarely know exactly what others earn, but we can see their mansions and penthouses and know their value from estate agents' windows. In addition, housing is, at least in part, a positional good in that the gains in well-being from its consumption depend on the attitudes and consumption decisions of others. As we argued in Chapter 5, our house reflects our identity and our standing, both in our own eyes and that of others. Therefore, satisfaction with our house is dependent on us not feeling inferiority or shame at our situation. A large disparity in housing outcomes is likely to reduce overall societal well-being as the possibilities of feeling shame are greater for those who perceive themselves as not meeting the societal norm. Therefore, inequality is important by itself.

Different groups in society may well be judged by governments according to their place in a moral order which will depend on political philosophy. Some groups may be thought to be more deserving than others, perhaps based on their relationship to the labour market or their position in a hierarchy. In some countries, ethnicity may also be an important factor. Governments may also vary in their attitudes to welfare and to work incentives. All of these factors may be reflected in the housing regime.

The importance of housing costs to the household budgets of many people means that changes may impact on the wider economy. Therefore, governments may want to intervene in housing to stimulate the economy by either increasing purchasing power or stimulating the construction of new build housing to increase employment. In contrast, when the economy is perceived to be overheating, housing may be used to slow down activity. For example, increases in interest rates will increase costs to mortgage holders and so are likely to reduce household expenditure on other goods and so reducing overall purchasing power in the economy.

Types of intervention

In the UK, the Barker report (Barker, 2004), which was commissioned to examine problems of affordability in housing, focused particularly on the issues surrounding housing supply and the perceived shortage of housing that was said to have resulted in higher prices. The major problem was perceived to be the lack of an appropriate level of response from housing developers to increases in housing demand that led to shortage and a concomitant increase in the prices that could be charged. Interestingly, most advocates of this position see government regulation through the planning system as being the major cause of the supply shortfall. If this diagnosis is accepted, then the primary focus of governments in dealing with affordability should be on the supply system, which was covered in the previous chapter, and specifically in reducing government intervention.

However, others have stressed the importance of demand-side factors in generating house price increases (see Matlack and Vigdor, 2008). These may range from general fiscal policies to particular factors such as the distribution of income between different groups and specifically the actions of the highest income groups in increasing their housing consumption and squeezing out others. From this perspective, interventions to deal with problems of affordability are usually financial as one would expect, although other types of intervention to help low-income households (such as the direct provision of state housing) are considered in other chapters. Issues concerning the general financial structure and regulation of housing were discussed in Chapter 4 and so the focus here will be on taxation and subsidy measures to deal with specific affordability issues and these will be considered in turn.

Taxes on houses can be applied in different ways. For example, they may be applied at the building stage, through taxes on materials, such as value added tax (VAT) or on land (such as through a land value tax); or through the appropriation of development gains in land, through planning obligations or a development tax; or on the consumption of housing, through a tax on the imputed rental income or a property tax; or on the sale or purchase of housing, through a purchase tax (or stamp duty in England) or through a sales tax; or on the increase in value through a capital gains tax which is similar to a sales tax as it is usually paid on the sale of property, or an inheritance tax payable on death. The justifications for housing taxation often relate to the need to raise government revenue; the need to create a level playing field with the treatment of other assets or revenues; or the need to make the housing market function more efficiently or effectively by, for example, offsetting volatility. Taxes on the increase in value of land or of housing are in place in some countries, but in varying ways. They are justified as a way of reducing house price appreciation and so easing long-run affordability problems and promoting the view of housing as a need or right rather than a marketed commodity. These taxes also reduce the returns from investment in housing and so play a part in reducing rentiership as well as lessening the processes reinforcing inequalities of wealth. Therefore, as we shall see in Chapter 12, the imposition of taxes on increased value is an important part of proposals to reform or replace the neoliberal housing regime.

Yates (2012, p. 398) defines subsidies as 'payments or financial aid given by the government to individuals or groups'. However, this definition covers what Yates terms 'explicit' subsidies that appear in government accounts such as housing allowances paid to consumers, deposit assistance paid to purchasers or direct grants to social housing producers. However, subsidies also usually include 'implicit' subsidies that do not involve explicit budgetary outlays or appear in government accounts. Examples include tax concessions to producers and consumers of housing and the setting of below market rents either in public housing based on cost or in private rental because of rent controls, all of which result in certain individuals or groups being treated more favourably than would have been the case without government intervention. The most extensive of this latter category of subsidies is tax concessions to owner-occupiers through allowances (reliefs) on mortgage interest payments and on relief from capital taxes. Subsidies may also be on the demand side (i.e. to consumers) or on the supply side (i.e. to producers). There has been a considerable shift in the balance between these in many countries over time. Yates (2012) states that supply-side subsidies in the form of direct government grants for the provision of public housing were dominant during the 1950s and 60s, but during the 1970s and 1980s there was a switch towards increased use of explicit demand-side subsidies, primarily in the form of housing allowances to individuals. However, since then there has been an increasing awareness of the weaknesses of a narrow approach on one type or another because of the growing awareness of the specific strengths and weaknesses of each and a growing consensus that a mixture of each was required in most circumstances.

If housing is thought to not be affordable then intervention can be on the supply or demand side or both. On the supply side, there may be many interventions that were considered in Chapter 7, one of which is subsidy to reduce the costs of the housing produced. Supply-side subsidies could be grants paid to local authorities, housing associations or private developers to produce housing at below market value. In the UK, planning regulation can be used by local authorities to require private sector developers to provide a proportion of any development at an 'affordable' level or, more accurately, below market price. The interventions may be general and apply to all new or existing housing, or they may be tailored to particular forms of housing aimed at particular groups who may be perceived as having an affordability problem, such as those having problems in accessing owner-occupation.

The strength of supply-side intervention is that it has a direct impact on the production and price of housing, and it avoids many of the weaknesses of the demand-side approach, but it can be difficult to target on those most in need without a complex system for determining access to the accommodation that takes into account changing circumstances. This explains the recurrent demand among some politicians in the UK (and in many other countries) that council housing rents should increase as incomes of the tenants rise and the extreme argument that tenants should be forced to leave if their income increases beyond a certain point.

The increased targeting of support on those in most need or on the lowest incomes has the drawback that it is difficult to avoid incentive problems such as the poverty and employment traps. These engage when income rises or employment is gained and a large proportion of the extra income is lost in withdrawn benefits and/or tax increases. This reduces the financial incentive to work or to work longer hours, although many other factors may be involved in individual decisions on these issues.

Therefore, although it is possible to target subsidy more effectively on those in most need using a demand-side approach, the disincentive effects of this can be a major weakness. Demand-side interventions also assume that they will bring about an appropriate supply-side response. For example, support to help marginal households become owner-occupiers assumes that the increase in demand thus engendered will result in more properties being built. But we have argued earlier that the supply of housing is inelastic, in that supply often responds to changes in demand after a long time-lag and then only partially. Therefore, both the short- and the long-term response to an increase in demand-side subsidies could be an increase in house prices that could feed through to increases in land prices as developers can afford to pay more for land. The same argument could be used for housing allowance schemes for renters that could result in an increase in demand that could be capitalised into rents and in to house prices that landlords are prepared to pay for their stock.

The weaknesses of each kind of intervention mean that many governments choose a combination of the two in an attempt to dilute the impact that one pursued entirely on its own would have, such as the disincentive effects of demand-side subsidies. However, the growing influence of the neoliberal ideology has led to the general reduction in supply-side subsidies in many countries with a consequent increase in demand-side subsidies such as rental allowance schemes or mortgage support with a consequent increasing trend in house prices in many countries.

There are a number of types of demand-side subsidies that are used in different countries. The first category is housing allowance schemes that provide support with rental payments to low-income households. Case Study 8.1 gives the example of rental assistance in Australia, which is in the form of payments to low-income renters in private rental housing and is much criticised for its level of support.

Case Study 8.1   Australian rent assistance

Labor federal governments turned increasingly to Rent Assistance (RA) as the policy instrument to assist people on pensions and welfare benefits to afford private rental housing. RA had started life as a small flat-rate supplement to the single aged and invalid pensions in the late 1950s but over time was extended incrementally to cover other pensioners and beneficiaries, with payments based on household type and size. By the late 1980s, further incremental changes extended the payment to families with dependent children in receipt of family payments and changes were made to the payment formula to target households paying higher rents as part of a strategy to reduce poverty, particularly among families with children. Subsequent changes were relatively minor adjustments, most designed to contain expenditures, such as introduction of a lower 'sharers' rate' in the early 1990s. There have been a number of failed attempts to reform RA. In the mid-1990s, the federal Labor government proposed that it be responsible for a new unified system of 'housing assistance payments' to all low-income renters (public and private) in which RA for private tenants would be 'substantially increased' and there would be protection for public housing tenants. These reforms attracted substantial criticisms from supporters of public housing and were unable to get agreement from the states and territories, subsequently stalling after a change of federal government. Coalition federal governments (1996–2007) indicated some disquiet about the mounting cost of the payment to the government and instigated reviews of RA but little change to policy settings resulted. More recently, there were three further attempts to address 'the problem' of RA under the federal Labor Government (2007–2013).

The proposal to pay housing assistance payments to the states for their public housing tenants as well as payments to private tenants was revived in the context of negotiations for a new National Affordable Housing Agreement in 2008. The Harmer

Pension Review proposed to restructure RA to further target those paying higher rents as well as 'addressing inequities that have arisen with the sharers rate of Rent Assistance' (Harmer, 2009, p. 94). The Henry Tax Review recommended that rates of RA should be substantially increased and linked to movement in market rents and that public housing rent concessions should be replaced by Rent Assistance and a new form of assistance for high-needs tenants, to improve equity and work incentives (Henry et al., 2009, p. 66). None of these suggestions for major reform were adopted, indicating policy inertia for almost 25 years, during which time more lower income households than ever became dependent on the private rental market. By June 2013, 1.27m 'income units' on pensions/benefits and family payments were in receipt of RA with annual expenditure of $3.63 billion, far in excess of that spent on public rental housing and other forms of social housing. It has been clear since the work of the National Housing Strategy in the early 1990s that lower income households in receipt of statutory incomes form the majority of those who are in rental stress in the private rental sector notwithstanding receipt of RA and that this remained the case in the 2000s (Yates and Milligan, 2007). As rents have increased due to high demand particularly in large cities, regular indexing of RA payments has become increasingly ineffective in enabling lower income households to rent affordable housing (Colic-Piesker et al., 2010). There is clear evidence over many years that RA, whilst it provides additional money to some low-income households, does not make housing affordable for many recipients and there is little evidence on whether and how they are able to sustain their tenancies. There is, however, clear evidence that RA has not stimulated an increase in the supply of lower rent housing (Hulse et al., 2014).

Source: Based on Hulse and Burke (2016).

Case Study 8.2 shows the example of a different system in the USA that is based on the provision of housing vouchers to some low-income households. Unlike many other forms of support, vouchers are limited, and not all eligible households receive them. In addition, there are problems with the involvement of private landlords in the scheme.

Case Study 8.2 Assistance for rental housing in the USA

Affordability

Housing Choice Vouchers (established in 1998 from the merging of two existing programs), enable low-income households to obtain housing that already exists in the private market. The availability of vouchers is rationed by the amount of funding allocated and so the program is not a 'right', and there are large queues of eligible households in some locations for vouchers. Moreover, having a voucher does not guarantee that low-income household will be able to use the subsidy (only 69 per cent succeeded in 2000). To succeed, the household must find an apartment or house that does not exceed the program's maximum allowable rent, that complies with the program's standards for physical adequacy, and whose owner is willing to participate in the program by agreeing to physical inspections which has proved to be easier in 'loose' as against 'tight' rental markets. They provide rental certificates to households up to 80 per cent of the area median and cover the difference between 30 per cent of adjusted family income and the assessed standard rent which is a proportion between 90 to 120 per cent of the 'fair market rent' which is 40th percentile of median rent and 50th in most expensive housing markets. Receipt of a voucher enables households to search in any neighbourhoods and the scope of destinations is wide. However, 'evidence shows that the vouchers have not succeeded in countering the forces of racial discrimination and segregation. Even though voucher holders have the ability to reside in middle-class neighbourhoods that are not racially segregated, most end up in predominantly minority neighbourhoods, most of which also struggle with high rates of poverty' (Schwartz, 2015, p. 241). This may be partly down to choice, as households may want to remain in a neighbourhood they know and minorities may want to reside in a neighbourhood where they feel safe and are not subject to discrimination. But constraints may also be important as most available properties are in low-income neighbourhoods.

In summary, vouchers are far less expensive per dwelling unit than project-based funding (such as public housing) and offer a greater degree of residential choice. However, the limited geographical spread of affordable housing units limits neighbourhood choice and cannot counter the forces of segregation. The approach is much less effective in 'tight' housing markets and some vulnerable households (large families, older people and individuals with health needs) tend to be less successful in finding accommodation.

Source: Based on Schwartz (2015).

The second category of demand-side subsidy is support for owner-occupiers. One widespread form of this is tax relief on the capital gains from the sale of housing by owner-occupiers which is common in a number of countries. It may be through tax relief on mortgage payments or lower than market interest rates on certain mortgages. Also, it may include specific payments to households to help them purchase housing. These are often focused on first-time buyers to enable them to access owner-occupation. An example of this is given in Case Study 8.3 of help for first-time buyers in Australia.

Case Study 8.3 Assistance for house purchase in Australia

First Home Ownership Grants (FHOGs) are paid directly to households purchasing their first dwelling, and supplement their savings and the mortgage obtained from a financial intermediary to meet the sale price of the dwelling. The Australian

Government established the first FHOG in 1964 as the Home Savings Grant Program. Since then FHOG programs with similar broad eligibility criteria, but with different names, have operated almost continuously. The Home Savings Grant was phased out from 1973 and replaced with the Home Deposit Assistance Grant in 1976. In 1983, it was replaced by the First Home Owners Assistance Scheme, which ran until 1990. The current FHOG was introduced in July 2000 to 'offset' the cost of the introduction of the Goods and Services Tax (GST).

This program history makes Australia a 'standout' country in the use of first home purchaser grants (Dalton, 2012). A further 'standout' feature of FHOGs in Australia is that they are not restricted by age or income, but are available to all who have not previously owned a dwelling. There have been four changes in FHOG program design. First, the idea that grants could be used to stimulate new housing supply was coupled to the earlier idea that grants would improve first homeowner affordability. The Australian Government made this explicit in March 2001 when it introduced the Commonwealth Additional Grant (CAG), which provided an extra $7000 to grant recipients who were purchasing a newly constructed dwelling (Dalton, 2012). Second, from 2000 the Australian Government transferred program administration to the states but with continued Australian Government funding. Third, some state governments added their own resources to these grant programs, while continuing with stamp duty concessions for first home purchasers. Fourth, some state governments also overlaid a spatial frame by providing additional amounts to householders buying in non-metropolitan regional areas, aimed at encouraging a more decentralised settlement pattern and addressing supply imbalances in metropolitan housing markets.

The conclusion from this analysis is that FHOGs do influence building activity, but only when additional money is available for new housing. Without this incentive, the effect of grants for existing dwellings on construction appears to be negative. Finally, there is the broader issue of housing supply where the evidence is that grants tied to the purchase of new housing increase supply but the effect does not last. Program evaluations suggest the programs stimulate a 'pull through', 'bring forward' or 'pull forward' effect, by encouraging purchasers to come into the market earlier than they otherwise would and, as noted above, convert intending purchasers of existing dwellings to switch to the purchase of a new dwelling. In other words, boost programs that stimulate new housing production do not address the underlying shortage in housing supply that has been well documented by the National Housing Supply Council. Nor do they necessarily address the inelasticities of housing supply that result in Australia having a housing system that responds more poorly than comparable countries to increases in house prices (Andrews, 2010, p. 7).

We however identified three features of FHOGs that show they fail as good policy. First, FHOGs for established housing undermine the objective of increasing housing construction. The modelling showed that increases in these grants were associated with less new construction. Second, the take up of FHOGs supporting new house building tends to result in an increase in outer suburban growth area building. This runs counter to policy emphasising increasing the density of metropolitan cities. Third, there is evidence that the house builders respond quickly by taking orders for new houses but then can struggle to maintain quality and complete houses on time. Finally, based on our modelling and other research, we noted that the link between FHOGs and housing price increases is unclear. However, it is clear that FHOGs do exacerbate the inequality in the underlying distribution of direct and indirect subsidies in the Australian housing system.

Source: Based on Taylor and Dalton (2016).

Supply-side subsidies are usually in the form of grants to public or private developers to provide housing aimed at reducing the rent or purchase cost to the dweller. One particular category is the cross-subsidisation by developers required through planning powers to provide 'affordable' housing. Other forms may be support in kind to the developer from a public body such as the provision of land at below market price or the use of compulsory purchase powers to enable site assembly which were noted in Chapter 7. Another type of supply-side subsidy is price or rent controls. Price controls on owner-occupied housing are relatively rare because of the political bias towards this tenure in many countries. More common are rent controls that may reduce the amount that landlords can charge in rent. Rent controls vary in their application. They can apply to all private rental properties in a country or apply only in specific locations or particular market segments. They may seek to keep rents at a specified level or allow increases for specific reasons such as general or house price inflation, or for actions by the landlord to improve the property. They may be for a limited period of time or permanent. They may apply to all properties or just those where tenants receive state benefits. The imposition of rent controls is usually combined with action to improve tenants' rights and to control tenancy conditions as well as measures to ensure that properties are kept in good condition (see Chapters 4 and 5).

Case Study 8.4 examines the rent-setting procedures in Sweden which has a form of 'soft rent control' in the public sector which has applied in the private sector as part of the 'unitary rental market'. The marketisation of the municipal housing companies has put this system under threat, and it is likely that market rents will gradually appear in the private sector.

Case Study 8.4 Rent setting in Sweden

In a market system, housing rents are set by the market and the market alone. This is not how rental pricing works in Sweden. Instead, pricing, for public and private rentals alike, is characterised by what Turner (1997a) has called 'a "soft" rent control system'. This system, today, has two main elements.

The first element is the so-called use-value (*bruksvärde*) approach. The crux of this approach is that rents should always be set according to a dwelling's use value, not its exchange (market-based) value. Utility-based rents were introduced in Sweden in the late 1950s in the public rental sector, before being extended to the private rental sector in 1968 (Hedman, 2008, p. 9). Previously, rents had been set according to individual-dwelling 'historic costs', whereby the rent charged should cover all historic capital costs plus ongoing operating costs such as maintenance charges – an approach which explains why private sector rents in the 1960s were often lowest in the most attractive, city-centre locations, where the housing stock was oldest (Kemeny, 1993).

Under the use-value system, the approach is different, although costs are still very relevant. Let us first consider the public rental sector (the reason for the choice of which will become apparent below). In setting rents, an assessment must first be made of the overall rental income required across a municipal housing company's portfolio to cover overall company costs, to provide a certain level of reserves and, from 2011, to generate a 'normal' rate of return. This cumulative 'necessary' rent can then be spread across the portfolio according, in theory, to the different housing units' individual use values (and once again, the reason for the caveat will be explained below) – above cost-covering rents being charged on dwellings with high perceived use values and low individual costs, with the surplus used to lower rents on properties with low use values and high individual costs.

The second element of Sweden's 'soft' rent control system concerns the linked question of how rents are actually set in practice. Again, the market is a somewhat distant chimera in this particular context. Rents are set instead through negotiation – between landlords (public or private) and the local tenants' unions (hyresgästföreningar) – based on the use value approach described above. The key point is that in practical terms rents are set not by the market but through what Ruonavaara (2012, p. 100) describes as a 'corporatist rent negotiation system'.

Until very recently, rent regulation in Sweden included a third 'pillar' alongside the above two elements; but this third element was abolished from the beginning of 2011 by new housing legislation which terminated the municipal housing companies' price-leadership role in the rental market. The essence of this role was that negotiated rents in the private sector had to be based upon rents in the public sector – hence why I started with the latter above. In generic terms, as Hedman (2008, p. 9) notes, the municipals' non-profit rents served 'as the first-hand norm within the entire rental sector', a role 'codified in law in 1974, after having been an implicit aim ever since . . . 1968, with the introduction of the utility value system'. The rationale, as Kemeny (1993) observed, was to prevent too large profits for private landlords. More specifically, private rents were 'not allowed to be significantly higher (5 per cent) than rents in similar apartments owned by the municipal housing company'. This regulation has now been discarded. In negotiations, private sector rents still have to be shown to be comparable to those for apartments of similar use-value, but the comparison should not only be made to apartments in the public sector, but also to those in the private sector.

It is possible that over time, the result of the recent removal of the third pillar of Swedish 'rent control' may be the beginning of a material divergence between public and private sector rents; and that this may be particularly notable in city-centre locations. Nevertheless, the negotiations-based rent 'comparability' principle (*hyresförhandlingslagens likhetsprincip*) still applies, albeit with the municipals' special role removed; as does the use-value system underpinning it. Any material shift in rents is thus likely to be slow – a likelihood engrained, in fact, in the new law, which includes a 'step' rule (*trappningsregel*) under which tenants subject to material rent increases (larger than about 10 per cent) can ask a rent tribunal to require their landlord to phase the increase over a number of years. Moreover, the power of the tenants' unions in negotiations – Ruonavaara (2012, p. 100) speaks of an 'exceptionally high degree of organisation in the tenant movement' – remains. All of which is to say that Sweden remains a long way from market rents and the uneven urban landscapes they tend to produce. As Bengtsson (1994, p. 185) writes, rent-setting in the private sector has constituted 'nothing but anticipated use-value comparisons based on the negotiated rents of the public sector', with the result that 'the system has all but lost its connection with the market'.

Source: Based on Christophers (2013).

The balance of taxation and subsidy and their application to different tenures are major factors in shaping the housing regime. Case Study 8.5 shows how they played out in Sweden and have changed from being 'tenure neutral' to a situation where owner-occupation is a favoured tenure, with many benefits to renters being reduced. This radical change has had profound impacts on the housing regime, but also has major implications for inequality and social justice between different income groups.

Case Study 8.5   Housing subsidies in Sweden

Tenure neutrality was a central premise of Sweden's post-war housing reforms, and received its most ardent expression in the government's Guidelines for Housing Policy (*Riktlinjer för bostadspolitiken*) established in 1974. Lennart Lundqvist (1987) has argued that this bill, and the policy regime it underpinned, emphasised three aspects of tenure neutrality: households in different tenures should ideally have equal standards, costs and influence (for example, over issues of housing management, maintenance and modernisation). However, the bill also sought to do something more subtle, but arguably much more significant: to accord equal social status to each tenure form, whereby no single type of tenure would be seen as more 'valid' or 'normal' than any other. In any event, the crux of the guidelines was that all and any policies enacted in the housing realm should be formulated explicitly in the light of the 'concern not to steer households to a certain housing form' (Lundqvist, 1987).

The situation today is a far cry from the ideals espoused in the 1974 guidelines. The reality is that the 'ideology of home ownership' examined by Richard Ronald (2008) is no less embedded in contemporary Swedish society, culture and politics than in the Anglo-Saxon and East Asian homeowner societies that Ronald focuses on. Equally significantly, and fundamentally bound up with this ideological transformation, tenure neutrality has long since effectively been renounced in policy practice and has been effected primarily through a series of separate fiscal and monetary adjustments during the past two decades that have entailed the state pulling back from its historic welfare role, while advancing, in tandem, a model of individual economic responsibility with which owner-occupancy sits most comfortably.

By any measure the most material policy adjustment has been the phasing out, since the early 1990s, of interest subsidies on new construction (Turner, 1999). To appreciate why this adjustment has had such significant implications for tenure equity issues it is imperative to understand that the system of subsidies in place up to 1993 actively favoured rented and tenant-owned housing over the much larger 'pure' owner-occupancy sector (which accounted, at that time, for about 40 per cent of dwellings). The mechanism by which it did so was as follows: the government set a 'guaranteed interest rate' for loans for new construction that was substantially lower for rented and tenant-owned housing than for owner-occupiers; subsidy was received by virtue of the government paying the difference between this guaranteed rate and the market interest rate. As Turner remarks, there was a clear tenure-equality logic to this approach, since 'owner-occupiers received a second subsidy through the tax benefits they received' (Turner, 1999) – benefits which we will discuss presently, and which, critically, have not been phased out. The scale of the impact of this monetary flexing was apparent as early as 1999, by which time the annual subsidy had declined to 7 billion crowns, from 36 billion just six years previously (Turner and Whitehead, 2002, p. 207) – although as Anders Lindbom (2001) notes, the decline was also influenced by falling interest rates and construction volumes.

The second set of policy measures to impact materially on tenure neutrality has been fiscal, and largely taxation-based. In the early 1990s, indirect taxes on rental and tenant-owned housing were increased (to fund a concurrent decrease in the marginal income tax rate for most wage earners) (Turner, 1997a, p. 480). More recently, in 2008, the property tax system was overhauled in such a way that owner-occupiers benefited disproportionately. A national tax – levied at 1 per cent of taxable value for single-family dwellings and 0.4 per cent for apartments (whether rented or tenant-owned) – was replaced by a municipal tax, levied at 0.75 per cent and 0.4 per cent, respectively. Moreover, the maximum payable was now capped at 6000 and 1200 crowns, respectively, equating to taxable values of 800,000 and 300,000 – a much more beneficial cap for homeowners than apartment owners/tenants, given that the prevailing national average taxable values were 975,000 crowns for single-family dwellings (that is more than 20 per cent above the threshold at which the cap kicked in) and 270,000 for apartments (10 per cent below the threshold). Notably, the government appears latterly to have recognised the tenure-tilting nature of these changes, deciding in its 2012 Spring Budget to lower the apartment tax to 0.3 per cent. Yet even this, public and private landlords claim, does little to redress the much larger taxation-related bias in the system: the fact that the continuing tax-deductibility of mortgage interest, a longstanding benefit that no Swedish administration has dared tackle, advantages individual home owners (including, significantly, owners of bostadsrätt) alone.

Third and finally, means-tested housing allowances (bostadsbidrag) have, over the past 15 years, been slashed. This development, driven by policy changes to limit eligibility (Turner and Whitehead 2002, p. 207), has not constituted a renunciation of tenure neutrality per se. Yet its effects have been overwhelmingly concentrated amongst rental tenants, where reliance on allowances is greatest, as opposed to amongst homeowners, thus making renting relatively more expensive; and it clearly constitutes a form of welfare state retrenchment explicitly located in the housing sphere. The numbers are stark: a 70 per cent fall in the number of households entitled to and claiming allowances between 1995 (576,000) and 2009 (174,000); and declines in total expenditure on housing allowances of about 40 per cent between 1995 and 1998 and a further 36 per cent between 1999 and 2008.

The cumulative impacts of the above three sets of policy adjustments – in interest subsidisation, property taxation and housing allowances – are apparent throughout the world of Swedish housing in the early twenty-first century. Two interrelated impacts can be usefully highlighted here, to underline the arguments about welfare retrenchment and tenure inequality respectively. First, as Lindbom (2001, p. 508) has observed, the housing sector went, in the course of just one decade, from imposing a net yearly cost on the state of 25–35 billion crowns (in the late 1980s) to delivering a net income of 31 billion (1999). And second, as higher construction and taxation expenses were passed on to renters, the cost of renting increased markedly not just in absolute terms but also relatively-speaking. Sven Bergenstråhle calculates that between 1986 and 2005, rents increased by 122 per cent, compared with a rise of 41 per cent in owner-occupancy costs and with general inflation of 49 per cent; other researchers suggest that something in the order of 70 (Jan Danneman, cited in Lindbom, 2001, p. 510) to 90 per cent of those rental cost increases were due to political/policy decisions. In other words, a society now ideologically wedded to the individualised and economically-responsibilist ethic of owner-occupation has seen the state increasingly withdraw from the rental sector, and the costs of inhabiting the latter increase accordingly.

Source: Based on Christophers (2013).

Although the precise pattern of taxes and subsidies varies between different countries, there are changes that can be associated with a neoliberal housing regime that the Swedish situation exemplifies. Two elements stand out. One is the change in relative balance from renting to owner-occupation with the latter becoming more favoured with tax reliefs and subsidies. This is associated with a reduction in subsidies to renting in general in Sweden, although in the UK subsidy levels have risen as rents have increased considerably in the dominant private sector, thus leading to increases in the total expenditure on housing allowances, despite attempts to implement reductions in payments. The second and related trend is the move from supply-side to demand-side subsidies. In the public rented sector, rents have increased reflecting the removal of supply-side subsidies, and housing allowances have increased in importance. In owner-occupation, tax reliefs and demand-side subsidies have grown, and some supply-side subsidies have been introduced under the guise of incentives to new production.

Evaluating the impact of housing subsidies and taxation

The Mirrlees Committee in the UK (Mirrlees and Adam, 2011) laid out four principles for a good tax. It should:

1 have as little impact as possible on welfare and economic efficiency;
2 should not be unduly costly to administer or comply with;
3 should be fair in procedure, discrimination between groups and have regard to legitimate expectations; and
4 should be transparent – people should be able to understand the tax system.

Barker (2014, p. 58) has added a useful list of aims of taxation on housing:

- to reduce the incentive for households to speculate on rising prices and so reduce both house price volatility and possibly the level of prices;
- to move towards a more equal distribution of wealth;
- to move towards a better distribution of housing space;
- to achieve a level playing field between different tenure types;
- to achieve a level playing field between housing and other assets;
- to stimulate more housebuilding; and
- to support environmental goals.

These are a wide range of aims and can be applied to subsidy as well as taxation and show how these two elements are an integral part of the achievement of government objectives in housing, although this list of aims would not receive agreement from some politicians and therefore will only be applicable in certain housing regimes. However, it is worth emphasising here how the taxation of capital gains in housing, including that on the main home, would help to achieve all of these aims, which is why it is being increasingly advocated by many in housing (for the UK, see Barker, 2014; Bowie, 2017). The key element of this form of taxation though is that it reinforces the view that housing should be a consumption good rather than an investment. Many problems in housing markets can be put down to the emphasis on wealth creation and capital gain that increases house prices, the volatility of housing markets as well as more distributional concerns about inequalities in wealth.

Yates (2012) provides an overview of the evidence on the impact of different forms of subsidy on affordability objectives.

> This evidence suggests demand-side subsidies are generally more effective in meeting affordability objectives than are supply-side subsidies because they can be well targeted (providing support to only those who need it) and are more cost-effective than supply-side subsidies in terms of their impact on government budgets. However, they are less effective in tight housing markets (where vacancy rates are low) and in housing markets where the elasticity of affordable housing supply is low. In such cases, they could increase the cost of housing in the affordable segment of the housing market and result in a net loss of stock affordable for low-income households.
>
> (Yates, 2012, p. 407)

There has been substantial criticism of the impact of demand-side subsidies aimed specifically at helping marginal owner-occupiers into that tenure. Again, it is worth quoting in full from Yates' review of the evidence.

> Demand-side subsidies that facilitate access are often seen to be inequitable (in that they provide support to households with fewer affordability problems and more choices than many of those in the rental market), inefficient (in that they simply add to price pressures in the housing market) and ineffective (in that they merely bring forward purchases that would have occurred in any case). There are also concerns about encouraging vulnerable households into homeownership and exposing them to the associated risks may not be in their best interests.
> (Yates, 2012, p. 409)

This general picture is mirrored in the situation in Australia outlined in Case Study 8.3. As Yates argues, there is little evidence that demand-side subsidies to owner-occupation increase the supply of new dwellings in that tenure. As was noted in Chapter 7, the supply system is very complex and development agencies may not be stimulated by increased demand, but may just increase their prices and pass this on in the price they are prepared to pay for future land purchases. Programmes designed to provide subsidy to enable households to enter owner-occupation may help the recipients in the short-term, but add to general price and affordability pressures for others and for similar households in the future.

General tax subsidies to owner-occupation are criticised for a number of reasons. The first is that they have resulted in higher house prices as the effective demand moves through to land prices. The second is that they tend to be pro-cyclical and so contribute to the boom-bust cycle in housing and make government management of the economy more difficult by counter-acting the effectiveness of automatic stabilisers. Third, they have been criticised for their distributional outcomes as they tend to benefit those with high incomes. General subsidies (rather than those focused on marginal owner-occupiers) tend to support people on high incomes whilst making more difficult (because of their impact on house prices) to enable moderate income households to enter the sector.

The impact of demand-side subsidies in rental housing is mixed. Although there has been concern about their incentive effects, Yates (2012) concludes from a review of studies that there is little evidence of any impact of housing support on labour market incentives. Nevertheless, housing allowance schemes are generally costly to deliver and difficult for renters to cope with. There is little evidence that demand-side subsidies increase the supply of new rental housing (see Case Study 8.1). In neoliberal housing regimes, there has been a phasing out of supply-side subsidies and a move towards removal of rent controls, and so rents have moved to market levels, thus increasing the financial strain on low-income households and housing allowances, resulting in efforts to reduce government expenditure by restricting availability. The example of Sweden in Case Study 8.5 illustrates this.

An alternative form of subsidy to rental sectors is supply-side subsidy through rent control, a major advantage of which for governments is that they do not bear the cost of the subsidy to the tenant. The direct incidence is on the landlord, but there may be further impacts as this reduces the price that landlords can pay for housing whilst still making a return on their investment. In turn, this will have a downward effect on general house prices and, therefore, on the price of land. The incidence is important as, if the impact is largely on landlords and they are unable to pass this on, it will reduce their income and ability to keep property in good condition. The main disadvantage of rent control is that it is perceived as restricting the supply of rented housing. It is argued that rent control in the UK over a long period of time has resulted in a reduction in the size of the private rented sector because of a lack of new supply, and this case is reinforced by the large increase in the tenure that has occurred in the UK since deregulation in the 1990s. However, an alternative interpretation of the UK experience is that the decline in size of the private rented sector occurred at a time when there was widespread provision of low-cost state housing, and owner-occupation was easily accessible, as well as the existence of rent controls. The recent growth in the sector has occurred at a time when state housing is in short supply, and there are affordability problems for many people seeking to access owner-occupation, as well as rent decontrol. The restriction or not of rents is only one part of a complex situation of interdependencies between tenures. Clearly, the imposition of rent controls needs to be implemented with a clear understanding of the context within which it will operate and the impact it will have across the housing regime. We will return to this in Chapter 12.

The interventions discussed in this chapter are designed to change the distribution of housing, and this is how they should primarily be judged. The key issue is whether each intervention achieves the distributional aims of the government that enacts it at the least cost when compared to other forms of intervention. But all of the interventions have other impacts such as the possible disincentive effects or the price impacts of some demand-side subsidies. Also, it can be difficult to assess the impact of any intervention because of its eventual incidence. For example, subsidies to households may first be given to them, but if they result in higher house and land prices the real beneficiary is the landowner. But the impact of these interventions is broader because of the impact over time and the fact that they are broader than just the housing outcomes as they may impact on the economy and on societal well-being.

Conclusion

In a market situation, the distribution of housing is likely to mirror the distribution of income and wealth in the society. However, housing is also a key driver of inequalities because of its high cost in household budgets and its mechanism for some people to derive wealth from the increase in house prices. Housing is a major part of the 'rentier economy' (Standing, 2017) that has been the consequence of the growth of neoliberalism. Housing subsidies and taxes can alleviate or support these inequalities. The neoliberal trend has been to remove general supply-side subsidies to public renting and to rely on demand-side subsidies such as housing allowances for renters that are focused on lower-income households, whilst retaining and improving the financial advantages of owner-occupation and offering help to some groups perceived as having affordability problem and encountering difficulty in entering the sector. The result is a complex picture of winners and losers amongst different groups, but, in general, a situation that may help to reduce poverty, but does little to reduce the growing inequality. Support for owner-occupiers in terms of tax exemptions, as well as specific affordability programmes, serves to reinforce the drivers of inequality through increasing house and land prices, whilst not bringing forward the increases in supply that would counteract this (see Chapter 7). Also, it creates the intergenerational inequities that have seen many young people unable to access owner-occupation, having seen their parents generate large wealth through the increase in house prices.

The policy tools are available for governments to pursue a more egalitarian policy. The imposition of taxes on the increase in house prices would reduce rentier income (which would benefit the economy in general as well as helping to reduce inequality) and put a constraining influence on house prices which would reduce affordability problems. The greater reliance of supply-side subsidies for public renting and the imposition of rent control as pertained in Sweden until the 1990s (see Case Study 8.5) would help low-income families whilst reducing government expenditure on demand-side subsidies and restraining house price growth.

Chapter 9

State-provided housing

Most of the other chapters have been organised around categories of housing problem and policy objectives, so why devote a chapter to a form of intervention such as state-provided housing? There are three main reasons. The first is that state-provided housing has become a key problem in many countries as, during the neoliberal era, it has become politically popular to downscale the sector by selling off the assets through right-to-buy or regeneration projects (see Case Study 6.1) and prohibiting new building in the tenure. Financial restrictions have also led to maintenance and management problems in some places and the residual nature of the remaining stock has fuelled stigma and identity problems among tenants. So, state-provided housing is increasingly seen as a problem to be dealt with by governments. The second reason is that the state provision of housing is a key policy that can help to achieve a wide range of housing objectives whether around supply deficiencies, affordability, house condition or egalitarian and sustainability issues. Therefore, its discussion is justified because of its importance as a policy tool. The third reason is that state-provided housing has become heavily politically charged and its demise in favour of owner-occupation has been a touchstone of neoliberal housing regimes. This may, in part, be a reaction to the symbolism that social-democratic regimes have given to the tenure, and its provision has, historically, been a major demand from trades unions and other labour organisations. As a consequence, the extent and importance of state-provided housing is often seen as a touchstone of the housing regime. The form, functioning and status of state-provided housing has been changed fundamentally by the neoliberal trends in housing in many countries.

Governments have provided housing themselves or through different agencies such as local authorities or housing associations, or housing banks. The motives may be to deal with problems in the supply system, and so it may be seen as a way of increasing supply. Or it may be seen as a mode of intervention to enable low-income people to access decent housing. Usually this form of housing is subsidised, although to varying degrees and in different ways. Sometimes the label 'social housing' is used to describe this sector, and this illustrates one of its possible roles in providing for those households who may have difficulty in securing decent housing in the market. However, state-provided housing has not always been aimed at the poorest of the population as the examples in the chapter will show.

The chapter starts by defining what is meant here by 'state-provided housing' and then examines the main objectives held for it and the roles it has performed in different countries and at different times. It will describe its main forms and the development processes involved as well as the management of the state housing stock and the different styles employed. Finally, the advantages and disadvantages of this form of state intervention are analysed and the findings compared with other forms of intervention.

What is state-provided housing?

The term 'state-provided housing' is not one in general use, and the labels and precise forms will vary from country to country, usually depending on the objectives of provision and the agencies involved in its supply and management. It is used here in preference to the term 'social housing' that is ideologically loaded and presumes a role for state housing only catering for the poorest of the population, whereas the sector may have a much more extensive role than that. 'Public housing' is a preferable term, but it has also developed a negative loading, because of the strength of the neoliberal attack. Therefore, the more neutral and accurate term 'state-provided housing' is used here, although it is not a term that is easy to define. It usually involves housing that is developed and/or managed or owned by the state at a local or national level, or by a state agency, and rented out to tenants selected on the basis of a set of criteria that may vary over time and place. A state agency may be difficult to define as many agencies have a mix of public and private elements. A good example is British Housing Associations that are legally private, voluntary, non-profit companies, some of which are charities. They are formally controlled by a Management Board elected by their members; however, these associations gain much of their funding from central government and are regulated by a central government agency (the Homes and Communities Agency in England). The regulation is very tight so that government can essentially control the operations of the associations in areas such as the type of housing that can be built (both its built form and its tenure and its occupiers) and the rents or prices that can be charged, as well as the main management policies and the type of lease that can be offered. The extent of the regulation led the Associations to be categorised as public agencies by the Office for National Statistics (ONS) although this categorisation has since been reversed so that they do not feature in public sector accounting. The key element in the classification is a judgement of the extent of state control over the provision and management of the housing. Of course, Britain also has the mechanism of council housing which is developed and managed by local authorities and partly financed and controlled by central government. However, even here there may be private agencies involved in the finance and building of the housing and even sometimes in its management. Therefore, the state provision of housing is rarely an exclusively state function, as the supply system may involve other private or voluntary bodies in some capacity as in the British case. Nevertheless, both council and housing association housing in the UK can be categorised as being state-provided housing because of the extent of state control of the process.

Other countries have different mechanisms of state provision. For example, Sweden operates through Municipal Housing Companies which are legally private companies, but are controlled by the municipal authorities (see Case Study 9.5). The situations in Australia, the USA and China are given in the Case Studies 9.1, 9.2 and 9.3 later in the chapter. The one country covered in this book that does not have a state-provided rental sector is Argentina. State help for low-income people here has been focused mainly on entrance to owner-occupation as high rates of inflation (35 per cent in 2017) have meant that rental is seen by dwellers as poor value for money as rental costs will rise in line with inflation, but will not be reflected in capital gain on house price for the dweller. In other words, ownership of a house is seen as a hedge against inflation by many people, whereas renting is seen as a commitment to ever-increasing costs. Nevertheless, this has meant that the lowest-income group have not been helped by the state, and so there is substantial government interest in the development of a state sector, although one where dwellers can enjoy some financial benefits of ownership through shared-ownership mechanisms or collective ownership through a community land trust. This example, shows that state-provided housing does not necessarily have to be rental, but could involve different forms of part or collective ownership (although full ownership would represent support for owner-occupation and so would not qualify under the definition used here). We will return to this later in the chapter.

The objectives and roles of state-provided housing?

State-provided housing has a long history in many countries, and the motives for its development have varied over time. In the UK, the first state housing was provided in the latter half of the 19th century by local authorities as an adjunct to their slum clearance activities. After the First World War in 1919, when the predominant tenure was private renting, which was subject to rent control implemented during the war as rents were increasing, the state saw itself as the most likely agent to deliver the large number of houses required to provide the 'homes fit for heroes to live in' for returning servicemen. As a consequence, the houses provided were of high quality and aimed at the skilled working class and lower middle class, rather than the poorest. However, in the 1930s when the acute shortage was assuaged and slum clearance restarted and the growth of owner-occupation meant that the middle classes were being provided for by that tenure, then the state focused on the poorest of the population and, in particular, on those displaced through slum clearance activity. As a consequence, the housing produced was of a lower standard with lower rental costs to improve affordability for low-income households and a rent allowance scheme to pay the costs of the poorest. The housing produced during the 1920s is still amongst the most popular of council housing and that built in the 1930s is largely stigmatised and unpopular, and many estates have been redeveloped or the subject of regeneration activity. This example illustrates two different roles that state-provided housing can play, and there are many more.

A common role of state-provided housing is to cater for those in most housing need such as those on low incomes or, because of other forms of disadvantage, find it difficult to access appropriate housing in the private sector and may experience homelessness. This role is usually given the title of *social housing*, which reflects its residual role in housing those who are not catered for in the market. Therefore, this role does not challenge market primacy. Case Study 9.1 of social housing in Australia provides an example of a small, residual sector that houses the poorest of the population.

Case Study 9.1   Social housing in Australia

Social housing in Australia accounts for just 4.7 per cent of total housing stock and represents the tenure of 'last resort' for some of the most disadvantaged individuals in society. This residualisation has been driven by the dual processes of budgetary constraint and increasingly focussed allocation policy (Jacobs et al., 2010). Historically social housing has been built and managed by the states with the bulk of funding provided by the Commonwealth government through agreements such as the Commonwealth-State Housing Agreement (CSHA) (1945–2008) and the National Affordable Housing Agreement (NAHA) (2009–present). The NAHA adopts a broader housing policy agenda with the objective that: 'all Australians have access to affordable, safe and sustainable housing that contributes to social and economic participation'. The NAHA also envisages a substantially increased role of the community housing sector in delivering affordable housing. While the most recent NAHA provided some new initiatives, there remains an undersupply of social housing. The nine years to 2007/2008 saw the national funding provided to Social Housing Authorities (SHAs) fall by 25 per cent in real terms (Pawson and Gilmour, 2010). Between 2001 and 2010 the total stock of public housing fell by 6.6 per cent, only partially offset by an increase in community housing. Nationally, in 2010 there were 383,316 social housing dwellings with a combined waiting list of 248,419 households (Shelter NSW, 2011). Together the reduced funding and focussed allocation has worked to define social housing as problematic in both policy rhetoric and public consciousness. The responses to these problems have been diverse, but have increasingly centred on large urban/estate regeneration projects (Arthurson, 2010a). At the centre of these projects sit two ideological policy presumptions. First, that 'mixing' social tenants with populations in other tenures can address the challenge of social housing (Arthurson, 2010b; Ruming et al., 2004). The second, and related, policy approach embedded in the urban/estate regeneration model is the use of private finance and developers to deliver change.

Source: Based on Ruming (2016).

Another role may be to meet the needs of residents in particular neighbourhoods or localities. The fundamental aim is to keep communities together and to develop and sustain community cohesion. This may be achieved by allocation policies that allow families and friends to remain in the same neighbourhood, through mechanisms such as giving precedence to people who can demonstrate a local connection over those from outside the locality with more housing need. The term *community housing* describes this role.

State-provided housing may be seen as a ladder to owner-occupation, enabling young households and others to have the low rents and security that enables them to prepare for access to owner-occupation. Sometimes this may be reflected in tenancy contracts that may be for short periods of time or may have stipulations that require relinquishing a tenancy if income rises above a certain level deemed to offer the opportunity of access to the private sector. This role may include the provision of forms of tenure other than rental, such as shared-ownership. This role may be termed *opportunity housing*.

A further role may be the provision of housing for people with 'special needs' that may involve different forms of housing than the mainstream or the provision of support services to enable households to sustain a tenancy and to undertake activities of daily living (Clapham, 2017). This role may be entitled *supported housing*.

State-provided housing may be aimed at specific employment needs such as for those defined as 'keyworkers'. An example of this is the provision of public rented housing in major cities in China to provide accommodation for migrants needed to meet employment demands. This role may be termed *worker housing*. Case Study 9.2 of state-provided housing in China shows how it has been used to house economic migrants to the cities of people unable to afford market prices.

Case Study 9.2 Public housing in China

The housing regime in China from 1949 to 1977 was part of the socialist system with a small owner-occupied sector and private housing transferred to local authorities. Therefore, there was widespread decommodification of housing. This system covered 90 percent of the urban population in the 1980s. But there was a key role for work units (danwei) in finance and production as well as management and allocation of the housing stock. The allocation of tenancies by councils and employers took into account need as well as civic or work contribution. Bricks and mortar subsidies were high and so rents were set at very low levels and were considered to be part of the 'social wage'. There were high levels of stratification in the regime because of the allocation by 'perceived merit' with bias towards high earners or the existence of corruption. People who were not in a work unit suffered problems in accessing housing and the resources of the work units differed with corresponding implications for the quality, quantity and cost of the housing provided.

With the liberalisation of the economy from 1978, work units became private and the state did not take up their housing role, but gave them and local authorities tax and land preferences to achieve housing goals. Work units were required to give housing allowances to their renters and to contribute towards a housing provident fund for their employees to buy housing. Various privatisation schemes were implemented to sell off existing public housing to tenants at large discounts and owner occupation was actively supported by the government.

This led to a regime from 1998 to 2008 that was characterised by a dominance of owner-occupation with rental housing (old and unsaleable work unit housing) restricted to the poor and disabled. So, state housing had a residual role in a neoliberal housing regime.

Since 2008 there has been a volatile real estate market with housing shortages and problems for migrants in access to housing in urban areas. The public rental sector has been revived and has become a core means of serving the housing demands of urban low-to-middle income residents, including both local residents and migrants. There has been the production of new public housing to sustain GDP growth and reduce housing shortages. The sector is the responsibility of local councils with strong central government directives.

Local government is responsible for the allocation of tenancies, which varies between different cities and property companies who are responsible for management and maintenance. The costs of maintenance are covered by rents, but if there is a deficit, local government may contribute. In response to directives from the central government, local governments have to take up all responsibilities for public housing development, including policy design, housing supply, location selection, land supply, facility management, provision of infrastructure, community services, allocation, and maintenance. In many cities, tenancies are set between 3 and 6 years. In Chongqing and Shenzhen, sitting tenants are allowed to purchase their public housing units after living in them more than 3–5 years. As the public sector was introduced only recently, there are still no clear policies on the level of purchase price and resale restrictions.

> In practice, rent levels in public housing were influenced and determined by both the market rent level and tenants' affordability. In most cities, it was set only slightly lower than the market rent level. According to the regulations in Beijing, the public housing rent is about 80 per cent of the market rent of nearby comparable housing. In Shanghai, the public housing rent is said to follow the so-called 'quasi-market rent' principle and the average rents of the first two municipal public housing projects are just slightly less than those of nearby comparable market renting housing. Therefore, affordability problems may arise among public sector tenants, especially for low-income households.
>
> Source: Based on Zhou and Ronald (2017).

Finally, any one sector in a particular country may attempt to fulfil a number of these roles in combination. However, there may be problems in this as the requirements of the tenure may be different in terms of the type of housing provided and the tenancy terms and conditions.

The provision of state housing can be aimed at almost all the objectives that the state may have in intervening in the housing market, whether the aim is improvement of the quality of the housing stock, the amount of housing built, affordability problems in general or for specific household groups or locations, issues of urban form and shape, sustainability and so on. Therefore, it is important in evaluating its impact to keep in mind the different and varied objectives that governments may have had in its development. A number of these objectives are described below and covered in more detail in the relevant chapter:

- It helps to increase the quantity and quality of housing produced.
- It provides a way of evening out cycles in housebuilding activity.
- If built with supply-side subsidies, it provides an affordable and secure home for low-income households.
- It offers an alternative to private renting and owner-occupation, thus helping to relieve pressure on those sectors.
- It provides downward pressure on rent levels as the state can take advantage of low-cost borrowing and economies of scale in construction.
- It can provide a mechanism of collective capital gain that can be used to fund new house-building at below market rates and to cross subsidise between expensive and cheaper building.
- It helps to reduce homelessness and provides a resource that can be used for schemes aimed at dealing with homelessness when it occurs such as Housing First.

The development of state housing

The main elements of a public supply process were considered in Chapter 7, and so only a few points will be covered here. However, it is an important area to consider because there may, in some contexts, be advantages in public over private provision that may be important in persuading governments to intervene in this way and have been influential in the growth of the sector in many countries at certain times. It must be borne in mind, that what is labelled 'public provision' may, in practice, involve a combination of public and private agencies, and the state may be involved at different stages of the process.

Therefore, the state can choose the level and stage of its involvement in the provision process. It could underwrite finance, provide a subsidy and leave the production process to private sector agencies or, alternatively, reach a public-private collaboration agreement to share functions and responsibilities. Choice of the type of involvement may depend on the objective to be achieved and the resources and capability of state agencies.

A major consideration for governments may be the control over the development that public agencies may be able to exert. It may be easier to determine factors such as the type and standard of building and the targeting of the consumers than having to persuade other private agencies to achieve your aims. In the UK, since its large-scale adoption in the 1920s, council housing has been at the forefront of improving the space standards of dwellings. Governments have used the sector to pioneer new forms of construction and to provide housing in places and on sites that may be problematic or unprofitable for private developers. The state may have advantages in site acquisition and assembly over their private counterparts as it may possess powers that are not given to private agencies. Also, there may be other advantages as they may be in a better situation to make use of publicly owned land and may be able to use public mechanisms such as compulsory purchase orders. In addition, it may be easier to co-ordinate infrastructure provision and the provision of facilities such as schools and shops when the planning agency and housing provider are in the same organisation, which may be the case in some state provision such as that provided by local councils. The possession by the state of both planning and housing powers may mean that land can be acquired more cheaply and planning gain may be captured and used to make the housing more affordable. However, often these functions can be located in different agencies and operate at different levels, so co-ordination is not always effective.

In terms of targeting, state-provided housing can take many forms and tenures and can be targeted at a wide range of households. Therefore, it can be aimed at specific access and affordability problems.

In addition, the state provision of housing may make the financing of housing supply easier and more efficient. The state has command of resources not available to private developers and can carry risk more effectively because of its size and ability to print money if there are problems. The state is regarded by investors as low risk and so can borrow at lower rates and over longer periods of time than private borrowers. Its financial strength means that it can use housing investment as a counter-cyclical tool in the management of the economy, thus helping to even out housing market and general economic volatility. State provision of housing also provides an effective mechanism for control of supply-side subsidies without the implementation of complex control mechanisms to ensure that subsidies are correctly used.

The management of state housing

Most state-provided housing has been rental, as this tenure is more affordable for low-income households. The provision of rental means that the state agency has the role of ongoing management of the property as landlord, unlike a developer for owner-occupation who may only have an interest in the property up to the first sale. Any landlord has a number of key activities that need to be undertaken if the rental is to be successful. Some of these are basic to the financial status of the landlord. Therefore, suitable tenants have to be found and assurances received of their ability and willingness to pay the rent. Rents have to be set for the property and collected. If the value and desirability of the property is to be sustained than repairs and maintenance have to be undertaken. In some cases, there may be a legal obligation to deal with faults that may harm the health or safety of the occupants and others, although the extent of these legal obligations, the ability of the different parties to ensure they are enforced and the sanctions associated with non-compliance may vary considerably between countries.

The above elements are the basic functions of the management of rental housing, but some landlords may move beyond this and undertake additional activities. This may be because they perceive this as being in their business interests by keeping their tenants or customers happy and attracting new tenants. Appreciation of the value of the property may be aided by a programme of improvements and proactive maintenance above that necessary just to maintain a liveable condition. But landlords may undertake activities beyond this if they have the interests of the tenant at heart.

The management of rental housing can be divided into two categories (see Caincross et al., 1997). The first has been labelled the 'contractual' approach which reflects the predominant approach in the private sector, with activities being undertaken that further business aims. The second category is labelled the 'social' approach and recognises the wider social aims that state organisations (or possibly other organisations) may have in the management of their housing. These social aims may be wide ranging and could include helping tenants into employment, improvement of housekeeping, employment or budgeting skills; help with health or social care needs; and many other functions that could improve the lives of tenants. At the same time, housing management may be used as a form of social control to change the behaviour of tenants or to reinforce behaviours deemed as acceptable such as taking up employment or 'respectable' lifestyles, such as abstinence from drugs or excessive amounts of alcohol. The key point is that the landlord function enables the state to use its landlord status to achieve wider social aims that may improve dweller well-being, although there is the danger that it may use these powers as a form of policing and social control that may reduce the well-being of tenants.

The powers and responsibilities vested in tenants in the tenancy agreement may vary considerably. If the sector is seen as a long-term destination, then tenants may be given security of occupation and a lifetime tenancy. If the sector is seen as a temporary place of residence for people in need, then tenants may be given time-limited tenancies, may have to undergo a probationary period to determine their eligibility and may be asked to move on or have their rent increased if their incomes rise during their tenancy.

State landlords may adopt practices of consulting with their tenants or enabling them to have a degree of control over the management and maintenance of the dwellings (see Caincross et al., 1997). This may be done for reasons of social justice as well as efficiency in attempting to make the service responsive to the tenants as consumers. Ideas in the management of state housing have changed along with ideas of public sector management in general with 'new public management' expecting public organisation to mimic the processes of private organisations and to use contractual relationships to undertake many of their activities. Examples include the requirement of Swedish Municipal Housing companies to act in a market-like manner (see Case Study 9.3), as well as the requirement on councils in the UK to subject their housing management services to 'Compulsory Competitive Tendering'. Many of these recent ideas of management are drawn from what is perceived as the best of private sector practice. When adopted, this approach sees the sector operate more like the private sector, which, in some ways, counteracts the need for it to be public. In contrast, the case for a public sector is strengthened if it operates in a distinct way from the private sector and, in particular, offers its tenants a different and superior experience.

The public ownership of rented housing raises the issue of the accountability of the landlord in a way that is not usual in the private sector. There can be democratic accountability through local councils or government agencies where they manage the properties. Other agencies, such as British Housing Associations, rely on government regulation and volunteer management boards for accountability. In addition, landlords may want to involve their tenants in the management of the dwellings. This may be done through feedback on service provision, through tenants' surveys or through tenant participation in decision-making. Processes and structures of tenant participation may vary (see Caincross et al., 1997 for a review) both in terms of the matters that may be considered and the extent of participation or control given to tenants. State provision of housing gives the opportunity for tenants to have much more collective influence over their housing, although in the neoliberal housing regime, the accountability to tenants has either been seen as being on a par with consumers of private sector services, or been downgraded through a complex web of contractual relations (see Grenfell Tower in Case Study 5.2).

Tenant control of state-provided housing may take many forms that could overlap with the voluntary or third-sector provision and include models of collective ownership and control such as co-operatives or community land trusts. The state can delegate ownership and control to dwellers if it chooses, and there are many models of this available.

Private and public rental

One of the key relationships is between state-provided rental housing and the private rental sector. Kemeny (1993) outlined two possible forms of this relationship, which he termed a unitary or dual rental housing market. In the latter, the state is differentiated from the private sector and is confined to a 'social role' catering only for the groups in most social need. This form of relationship is common in neoliberal housing regimes where the private sector dominates and the state sector is constrained so that it is unable to compete with the private sector. In some housing regimes (such as the USA), the private sector primacy is so advanced that the future of the state sector is in doubt and the private sector is seen as the tenure for even the lowest-income tenants. In the unitary rental market, the public sector dominates in matters such as the setting of rents and the rights and obligations of tenants, where there is parity between state and private tenants with regulation designed to enforce common conditions. Some elements of this system are still evident in Sweden, which was the source of Kemeny's category. One example is the setting of rents which is done through negotiations between the Municipal Housing Companies and the Tenants Unions with the results applying to both state and private sectors. Therefore, rent (and other forms of) private sector regulation are set by the situation in the state sector, although this position is changing (see Case Studies 8.5 and 9.3).

The fortunes of the state and private rented sectors are closely inter-twined. In the neoliberal housing regime in the UK, growth in private renting has occurred at the expense of decline in the state sector, and the deregulation of private tenancies has led to the reduction in rights for state tenants as the government does not wish to see a situation where the state sector has a competitive advantage over the private sector. The security of council tenants has been reduced as new tenants are subject to 1-year probationary tenancies that may be extended only if behaviour is deemed to be satisfactory. Also, government has replaced secure tenancies for life with fixed-term tenancies that may be subject to conditions. For example, government has voiced its wish to be able to end the tenancies of households where the income rises above a certain level. The overall intention seems to be to equalise the rights of public sector and private tenants by reducing the rights of the former in order to prevent state housing having a competitive advantage.

The concern with the challenge of the state-provided sector to the market has been associated with what Kemeny (1981) has termed the maturation of the sector. For example, council housing in the UK has been funded through borrowing and from government grants (see also the Swedish system in Case Study 8.5). As the sector matures the loans are repaid, and the value of the properties increases from the historical build cost if it is valued at a market level. This increase in value potentially allows councils to build new housing through loans secured on the value of the asset of the existing dwellings. At the same time, the rental costs can be reduced through cross-subsidisation with the older dwellings. This is possible if rents are set on the basis of the utility to tenants rather than the historic costs. In other words, the rent charged on houses of a certain type and size can be set as the same despite the different times of building and historic costs. This means that the rental costs of new building for tenants can be much lower than in the private rental sector where the housing stocks are smaller and are usually newer. It is at times when this competition with the private sector has become acute that privatisation policies have been pursued. The benefit of the surplus generated through historic housing activities by the state is then privatised and individualised in the form of a one-off bonus to individual tenants (through mechanisms such as the 'right-to-buy') rather than being held by the community to form the basis of future housebuilding.

In other housing regimes, state-provided housing may be considered a mainstream tenure that can provide for a large proportion of the population, and not necessarily confined to those from the lowest-income sections. The sector has historically achieved this role in the countries of the Soviet Union as well as in Scandinavian countries and in the UK in the 1960s and 1970s. However, the sector did not achieve popularity with the population at large, and so the political priority given to owner-occupation was perceived as a vote winner by politicians. Policies of privatisation of state housing and support for owner-occupation were popular. Owner-occupation was perceived as the tenure that gave status and control as well as a way of building wealth. As we shall see, in contrast, the state-provided sector was perceived as reflecting failure, and in some cases stigma, and offered a future of rental payments that did not lead to an accrual of wealth and continued beyond the life of a mortgage. State-provided housing did not come out well in the dominant neoliberal discourse.

The privatisation of state housing

In many countries, and in all but one of the countries covered here, the provision of new state housing has reduced substantially, and the sector is in decline both in terms of size and popularity with governments and public opinion. The exception to this overall picture is in China where large quantities of new housing are being provided by the state in many of the larger cities. However, in the UK the total proportion of state-provided housing has fallen from 31 per cent in 1971 to 18 per cent in 2013 (both figures include stock owned by housing associations). In England, councils have been prevented from providing much new housing by the financial controls of central government and housing associations have been restricted to providing houses for sale or at only 20 per cent below market rents. The Right-to-Buy policy remains in England (though not in Scotland and Wales and Northern Ireland), and about 12,000 houses were sold in 2015 (Wilcox et al., 2016).

The privatisation of state housing has been a feature in many countries since the rise of neoliberalism from the 1980s. Perhaps the most extreme example of this is in Eastern European countries after the political changes associated with the fall of the Berlin Wall and the loss of power of Russian-backed communist regimes. The traditional East European welfare regime had consisted of almost universal state ownership of all housing and state provision of new housing either through state-owned employers or through local authorities (see Clapham, 1995). In many countries, housing was seen as part of the social wage and so rents were very low (in Russia about 1 per cent of household income) and did not meet ongoing management and maintenance costs. As a consequence, much of the housing was in disrepair, and there was widespread criticism of the standard and form of new-built housing that usually consisted of high-rise apartment blocks in large estates. With the political changes, privatisation of housing was seen as a popular policy that saved public money (in foregone maintenance and management expenses and a capital gain from the price paid by households) and was a potent political symbol of the private economy that governments sought to institute. Privatisation took the form of restitution of property to previous owners who had seen their property taken over by the state in the communist era, as well as the sale of existing housing to the tenants at prices that varied between countries with some selling at a large discount and others at nearer to a market price (see Clapham et al., 1996, for a more detailed discussion). The privatisation policy was very popular in most countries and spearheaded a move towards widespread owner-occupation. But it had some deleterious impacts. The transfer of ownership of property in poor repair to low-income households reinforced the longstanding stock condition problem with some households becoming responsible for a deteriorating asset without the funds to undertake the necessary repairs. Also, in some countries, the privatisation of single apartments in large blocks was undertaken without a clear and workable mechanism for the maintenance of the common areas and the structural elements such as the roof and outside walls. The result has been the sometimes ad hoc maintenance activity.

In the UK, privatisation has largely taken the form of the influential and widely copied Right-to-Buy policy that has enabled existing tenants of councils to buy their property at a discounted price. The discount is dependent on the length of occupation of the tenant and has varied at different times. The argument for the price being set at below market price is that the price with a sitting tenant would be low, plus the argument that the tenants have already paid a large amount through their rent which, it is argued, should be seen as already contributing to the purchase. On the other hand, early research (for a review, see Murie, 2016) showed that purchasers tended to be the more affluent tenants living in the more popular properties, and so the implicit subsidy to them was regressive in distributional terms. In addition, the sell-off of the most desirable properties, that were often also those in the best condition, left local councils with responsibility for the property in need of repair whilst, as most of the proceeds of sale were captured by central government, they had few funds to undertake this, and so the condition of the remaining stock has been problematic, which in turn has stimulated recent regeneration initiatives that were considered in Chapter 6 (Case Study 6.1).

Later research (Murie, 2016) has shown that the longer-term impacts of the Right-to-Buy have been problematic in financial terms. The first buyers received a large subsidy in many cases when they sold into the open market after the sale restriction period elapsed. Later purchasers did not receive any discount and bought at market prices. A substantial proportion of properties (estimates are between 30–40 per cent; see Murie, 2016) have found their way into the private rented sector being let at market rents. Many of the tenants of these are receiving Local Housing Allowance because of the level of rents compared to their household income. Because council rents were much lower, Housing Benefit payments to the same tenant in the same property would have been less had they been retained in the council sector. Murie (2016) estimates that the addition to the Housing Benefit cost is over £800m per annum. Therefore, the long-term financial benefits of this form of privatisation have been limited. Despite these drawbacks, the policy has been popular in England and has been extended, in part, to include housing associations, although there have been moves to remove or restrict the policy in Scotland and Wales. The policy has been popular in a number of countries where it has been pursued such as former communist states in Eastern Europe.

In the USA, the federal government has instituted a number of programmes such as Hope VI to demolish state housing and to give the tenants Section 8 vouchers so that they can find housing in the private rented sector (see Chapter 6). This seems to be a popular policy with many parties including the tenants themselves (see Case Study 6.4).

Privatisation of state housing has taken many forms, and in Sweden, it has involved (among other ways) the changing role of the Municipal Housing Companies with the obligation placed on them to behave in a commercial fashion (see Case Study 9.3). This has changed their allocation procedures and the make-up of the sector with those most in need being increasingly housed through the mechanism of secondary tenancies.

Case Study 9.3 Municipal housing companies in Sweden

In the post-war period, Swedish public housing, made up of municipal housing companies (MHCs), has been an important instrument in the national housing policy to contribute to socially inclusive cities and municipalities. The socially inclusive undertaking of Swedish public housing has primarily been conveyed through its universal character, meaning that the MHCs provide rental apartments for the general public and their various needs (Bengtsson, 2001, 2004). Public housing has been directed to all, 'for the benefit of everyone'.

In a selective housing regime, the state is responsible for providing for households of lesser means alongside the general housing market. In contrast, a universal housing regime is dependent on the relation between the welfare state and the ordinary housing market in providing housing for the general public. A universal housing regime is characterised by the state being expected to correct the general housing market, so actors on the housing market can provide housing for all types of households, regardless of their economic situation (Bengtsson, 2001). While a selective housing regime links to, in particular, the liberal welfare regime in Esping-Andersen's typology, a universal regime could be clearly linked to the universal principles of the social democratic welfare regime.

Within public housing, gradual changes have made MHCs more market-oriented and subventions for public housing have been removed. In 2011, new legislation on Swedish public housing was introduced, making public housing a market actor. An important difference is that all actions of the MHCs must be economically justified; investments cannot be undertaken if they are not calculated to yield. However, the social responsibility of MHCs is still clearly stated in the legislation. This somewhat ambiguous mission has strengthened the view that MHCs are hybrid organisations (Christophers, 2013) in the sense that they not only should relate to business in a competitive market characterised by profit motives, but also have the aim of societal benefit. MHCs have reacted to this situation by raising the financial barriers to entry to the sector in order to reduce the risks of financial default. At the same time, there has been a large increase in the secondary housing market where local authority social services lease accommodation from the MHC and sub-let to low-income families.

The universal and inclusive approach of public housing in Sweden is changing as a consequence of gradual market adaptation, and in particular, as a result of legislation introduced in 2011. It could be argued that this shift can be described in terms of a transition from a universal model connected to the social democratic welfare regime to an ambiguous model with universal discourse, but with selective outcome. New Public Housing could be argued to be losing its inclusive character of 'for the benefit for everyone', and instead, is contributing to a more exclusionary rental system. The main arguments about the characteristics of New Public Housing could be summed up in the following three conclusions.

First, the thresholds for entering New Public Housing are increasingly excluding people with low or irregular income. Housing allowance (one of the state's interventions for building the foundation for the universal model) is often not accepted as legitimate income for a tenant to obtain a rental contract. Second, the raised thresholds are compensated by an increase in 'social contracts', where municipalities rent apartments from public housing companies, and in turn, sublet (under precarious conditions) to people who cannot gain access to housing through the regular market. This could explain how the increased residualisation of public housing appears in conjunction with raised financial thresholds. Thirdly and finally, this ambiguity means that New Public Housing is increasingly catering for the most well off and the most vulnerable in society as a consequence of the ongoing market adaptation. Public housing could claim inclusiveness and social responsibility, while at the same time, taking less economic risk. However, this is at the expense of a large group of people who become excluded from public housing or enter on restricted and temporary contracts with low probabilities of being transformed to ordinary rental contracts. While New Public Housing still clings on to a universal national discourse, the local outputs are clearly selective. The selectivity is, however, different from that of the traditional selective housing regime, as the target group is not only the most vulnerable, but also the financially stable middle class.

Source: Based on Grander (2017).

So why has privatisation of state housing been pursued so avidly in many countries? For countries in the neoliberal welfare regime, state-provided housing may be perceived as a challenge to the market and privatisation seen as a re-assertion of market relations. But many governments on the right of the political spectrum have, in the past, supported state housing, and few governments have argued for its complete removal. The political argument is usually over its role and whether it should be restricted to those who cannot access the private sector, playing a residual role. But here the question is the extent to which the private rented sector can take on the social role. Examples of homeless policy in the UK (see Chapter 10) and the example of Section 8 vouchers for public tenants in the USA, show that governments are attempting to test the boundaries of the sector by looking at ways that it can take over functions previously undertaken by the public sector with varying results.

Pros and cons of state housing

Many of the evaluations of state-provided housing are difficult to review objectively as they are so tied in with political discourses. Perhaps more than any other housing policy instrument, it has been imbued with wide political significance. The introduction of state housing in the UK was partly a result of its position as a key demand of the political left as well as a pragmatic response to housing shortage. Privatisation of state housing has been a potent symbol of a home-owning and a capitalist society as reflected after the demise of communism in Eastern Europe or the Right-to-Buy policy of the Thatcher administrations in the UK. The political attitudes towards state housing have also been reflected in the type of properties built with housing aimed at the poorest sections of the population, such as in the UK in the 1930s being of a poorer quality. Many of these estates have been problematic and unpopular, and many have been demolished.

The concept of stigma is at the heart of perceptions of state housing and ultimately its use as a mechanism of intervention in housing. In neoliberal regimes, where the scale of state housing has been small, and it has been given a residual role, catering largely or exclusively for the poorest sections of the population, then it has been stigmatised, resulting in unpopularity and lack of political support. The term 'social housing' encapsulates this role. The impact of this on tenants can be profound and cover issues such as employment prospects, health, self-esteem and general well-being. The case study of the USA (Case Study 9.4) shows how the residual role can emerge and the factors that have led to this situation. As Schwartz argues in the case study, this situation is not inevitable or irreversible. It is easy to see that further reduction in the size of the sector and its replacement by Section 8 vouchers would be popular with tenants and voters. However, the evidence is that tenants still end up with other low-income people in unpopular and low-affordance neighbourhoods, and so their situation does not improve markedly.

Case Study 9.4 Public housing in the USA

Public housing in the USA is regarded as being unpopular and problematic. Little new public housing is being built and many existing estates are being renewed through the Hope VI and other programmes with only a quarter of existing residents being rehoused and little public housing provided through the redevelopment. So, is this an indictment of the general concept of public housing? Schwartz (2015) argues not and that the faults in the US sector can be traced to the compromises in the original 1937 'New Deal' legislation that hindered its progress. He identifies four factors that have been important.

The first is tenant selection policy. To avoid opposition from the real estate industry it was agreed that public housing would not compete with the private sector and would only take the poorest of the population. As a consequence, there emerged a growing concentration of the poorest and most vulnerable households in the sector, with in 2013, 55 per cent with incomes of less than \$10,000 per annum (Schwartz, 2015).

The second factor is project location, with most public housing being situated in low-income and often minority neighbourhoods. This was mainly because the legislation that enabled the provision of public housing was not mandatory and so was taken up predominantly by cities and communities where need was most extreme and they lacked the power to provide housing outside their own boundaries. It was also because there was protest from many white neighbourhoods about public housing in their areas as well as black politicians wanting development in their areas.

The third factor, is the design and construction quality of the dwellings. As Schwartz (2015, p. 172) states:

> Public housing is almost always easily recognisable. Whether high rise or low rise, the physical appearance of many, if not most public housing projects differs sharply from the rest of the neighbouring housing stock. Public housing is usually built more densely, it is often isolated from the surrounding streetscape, and it is almost always assiduously devoid of ornamentation or amenity. The physical quality of the housing is frequently markedly inferior to that of other rental housing.

This was partly due to the severe cost constraints imposed on public developments, but also the aim to avoid competition with the private market. However, the result of the poor construction standards was often high maintenance costs.

The fourth factor is the poor standards of management adopted by many public housing authorities. Schwartz (2015, p. 175) argues: 'Although many public housing authorities have professional, highly competent managers, others have long histories of ineptitude, if not corruption'. There seems to have been little effective oversight of management procedures and processes and little accountability of the standards achieved.

The case study of US public housing does not prove the case against public housing in general, but it does show the problems that perhaps inevitably occur when a sector is provided on a low-cost, low-standard, segregated basis that provides fertile ground for stigmatisation. If they are to succeed, public housing systems need to be large, provide for households of different ethnic and socio-economic status, be in high-quality housing indistinguishable from the private sector and to be integrated spatially into the communities in which they are sited.

Source: Based on Schwartz (2015).

For social-democratic regimes, state housing has had a more positive political discourse, and at times and in particular countries, state housing has been a general tenure catering for a wide range of the population. For the future, it is necessary to examine this experience to see what are the possible advantages and drawbacks of this form of policy mechanism.

Status and stigma remain at the heart of a number of the criticisms made of state housing. For example, it could be argued that its provision has led to the segregation of low-income (and sometimes ethnic minority) households in low-status neighbourhoods, because of the development of much state housing in mono-tenure estates. It is undoubtedly true that much state housing has been provided in this way, but it does not have to be in the future. Although there may be economies of scale in both development and management of estates rather than 'pepper-potted' houses scattered throughout an area, this advantage is outweighed by the costs of the segregation to the households concerned and, it can be argued, to the society as a whole because of its impact on social solidarity. The experience in the UK of providing affordable state housing through the planning mechanism shows that it can bring real benefits in terms of reduced stigma, even if it does not seem to influence social contacts (see Chapter 6). A similar argument applies to criticism of the design and standards of state housing. Some state housing has been of very poor standard and has been unpopular and led to many problems of maintenance and repair, leading in some cases to early demolition. However, there are examples in a number of countries of high-standard dwellings being built and remaining popular. It can be argued that the lack of a profit motive in the public sector leads to the empowerment of designers, planners and architects to pursue 'innovative' designs and construction methods that have not always been successful. One example in the UK is the provision of high-rise and system built blocks in the 1970s (see Dunleavy, 1981) that have become unpopular and have caused considerable problems. But again, state housing does not have to be built in this form. The accountability of architects and other professionals can be countered by participation mechanisms that give accountability to tenants and prospective tenants and occupiers. It is salutary to remember that state-provided housing does not have to be rented housing, and so a variety of tenure forms could be developed including collective and individual forms of ownership and control that improved accountability and could counter concerns about stigma.

New collective form of ownership and control can also help to alleviate a major criticism of state rented housing relating to the lack of capital appreciation for the residents. An owner-occupier gets the benefit of any increase in the value of the property, and in recent times and in many countries, these gains have been substantial. A strong criticism of existing forms of state housing is that they do not enable tenants to enjoy the benefits of capital appreciation. Of course, their properties may increase in value, but the benefit is shared and can go to subsidise further construction and so help future tenants. Or it can be used for other social purposes. In other words, the benefit can be collectivised and socialised. However, in a society that rewards private property ownership, the lack of an asset such as an owner-occupied house can be important to the well-being of the individual. For example, the policy of 'asset-based welfare' has been influential in many countries. This is based on the concept that people build up wealth during their working lives that they then use to fund the income and services needed in old age. Of course, people who are unable to take up employment for any reason miss out on this opportunity as do all those unable to access ownership, and this could impact on their life chances and well-being in later life. The solution to this problem has to reside either in removing the benefit to all of capital appreciation by reducing the commodity element of housing, or in enabling tenants to enjoy these benefits also, or some mixture of both.

Another argument against state rented housing is that the rights of tenants over their house and their ability to control it are not appropriate when compared with the position of an owner-occupier. Repeated surveys have shown that, in societies where owner-occupation is a social norm, renters, whether in the public or private sectors, feel that they lack status and control over their house, and fewer people feel able to make a home in renting. In some societies, there is a difference between the public and private sectors, although as the examples of the USA and UK show (see Case Study 9.1) the relative position of the two sectors varies according to the status and conditions, both physical and in terms of rights and obligations, in the two sectors. In other countries where the ownership discourse is not so dominant, state-provided rental housing is seen by many people as a secure long-term home.

Conclusion: the place of state-provided housing

The chapter has shown the problems that have pursued state housing and its tenants in the neoliberal housing regimes. It has been regarded as a second-rate tenure that is only appropriate for those on the lowest incomes. In the UK, this 'social housing' has suffered poor conditions through lack of funds for maintenance (see Case Study 5.3 of Grenfell Tower in London) as well as restrictions of the rights of its tenants through short-term tenancies and increases in rent for those on above-minimum incomes and reductions in benefits for perceived over-consumption of space through the so-called 'bedroom tax'. However, the analysis of the US situation shows that many of the perceived limitations of state-provided housing could be avoided, and the example of China shows how the state can use the tenure effectively to meet particular housing problems (Case Study 9.2).

So, a key question is whether the unpopularity of state-provided housing in some countries can be overcome. Is there a political constituency for expansion of the sector and can the dominant discourse of owner-occupation be countered enough to make a popular case for the sector? The wide scope of the objectives that state-provided housing meets and the seriousness of the housing problems that have resulted from the neoliberal discourse offer scope for a more positive view of the worth of state-provided housing. Examples of particular issues have been covered in previous chapters such as those around the quantity of new housing supply which the resumption of state activity could help to rectify. Other issues are the affordability problems of young people as well as the perceived problems of high rents and poor conditions in the private rented sector. The example of China (Case Study 9.2) shows how state provision of housing can be justified on the pragmatic grounds of being the most effective way of achieving the objectives of increasing the supply of affordable housing for people considered to be necessary for best functioning of the economy. However, the provision of new state housing probably goes hand in hand with some of the other measures suggested in this book to restrain house prices through taxation reform that captures the investment returns from the ownership of housing and reinforces the concept of use value in housing, as well as regulation of the private rented sector to create a level playing field between the two sectors.

Chapter 10

Homelessness

Homelessness is included here because it is the most critical form of housing exclusion and can result in severe hardship for those people who suffer from it. Politically, homelessness can be an emotional and important topic that, when articulated through public opinion or the media, can put pressure on governments to react. The urgency and importance of the issue is related to its visibility. Although there are many other forms of homelessness, street sleeping in the middle of cities can be an uncomfortable and unpleasant reminder to many people of the problems of their society and paints a poor picture to visitors. But the phenomenon of homelessness also causes governments to confront and find answers to some of the fundamental issues concerning housing. Is housing primarily a market commodity or a fundamental human need, and can these two elements be reconciled? Do people have a right to shelter or some form of decent housing, or is it purely a matter for the market? When is someone homeless, and what constitutes shelter or the absence of 'home'? There are many answers to these questions, and so there have been many approaches to the definition of homelessness, the recognition and attribution of its causes and methods of dealing with it between countries and at different points in time. For example, homelessness in the UK was reduced substantially by the Labour governments in the years up to 2010, but has since increased with changes of government and the move towards a more neoliberal housing policy. There are variations in homelessness associated with the housing regimes in different countries, and the general housing policy adopted influences the extent and form of homelessness, as well as the solutions that a government will consider to be appropriate in dealing with it. Nevertheless, different approaches, such as 'Housing First' have emerged from a neoliberal country (USA), but have been adopted in many countries with different housing regimes (such as Sweden).

The chapter begins with a discussion of definitions of homelessness and its causes as there are many competing discourses that are politically contested. Two are considered here namely the 'minimalist' and 'maximalist' discourses. The chapter then focuses on government policies to deal with homelessness such as 'Housing First' that has been one of the most popular recent policy approaches to homelessness in many countries. This is compared with other approaches such as the 'staircase' model, and examples are given of the approaches in different countries. Finally, the chapter assesses the evidence on the impact of different approaches and the influence of different housing regimes on the incidence of homelessness and success in dealing with it.

The discourses of homelessness

There are different definitions of when someone is homeless. Fitzpatrick (2012) argues that there is a minimal definition of homelessness that receives almost universal agreement in the developed world. She labels this 'literal homelessness' and includes in it people who are sleeping rough and those in homeless shelters. Some countries move further in their definitions. For example, Sweden includes people living in some institutional settings as well as some who are living temporarily with friends or family. In Australia, homelessness is defined as having 'inadequate access to safe and secure housing', and in England, it includes all those without access to 'reasonable accommodation' to live in with their families. Terms such as 'reasonable' or 'safe and secure' show the arbitrary and difficult task of defining what is appropriate shelter. Sometimes the distinction is made between the condition of not having shelter and not having a home. However, this distinction is difficult to draw when it comes to defining 'shelter'. It is probably possible to universally agree that shelter has to be dry and warm, but does it have to be safe or comfortable? If so, how safe or comfortable as these things are relative? If safety is considered to be an element of shelter, then does it mean that a physical shelter is not enough if there is the threat of violence? Similarly, if a person feels insecure because of a threat of eviction, does this qualify as being a lack of shelter? One attempt to deconstruct the different elements that can be included in definitions of homelessness is provided by the European Typology on Homelessness and Social Exclusion (ETHOS) that has four categories namely: roofless (sleeping rough and in a shelter), houseless (people in homeless shelters or receiving support for homelessness), insecure (people in temporary accommodation or under threat of eviction or violence) or inadequate (people living in unfit or overcrowded or temporary situations). Speak (2004) argues that most definitions of homelessness are derived from the experience in developed countries and do not fit the situation in developing countries. She suggests a threefold categorisation of homelessness depending on the degree of choice or control that people may have. The first category is 'supplementation homelessness' where people move from the countryside to urban areas to increase their earnings, most of which they send back to their families. There may be sufficient income to afford shelter, but it is often foregone in order to maximise the amount sent home. People in this category do not make many social ties in their living environment as they perceive it to be temporary with their real home still being in the countryside. It could be argued that some migrants in developed countries could fall into this category, living on the streets or in overcrowded accommodation in order to maximise the money sent to their families in their home countries.

The second category is 'survival homelessness'. Here, as in the previous category, migrants have left home in order to gain an income, but in this case, they are unable to supplement their income sufficiently to send enough home to retain their position there and so become unable to return. Therefore, their homelessness is more permanent, and they may begin to make social ties in their new neighbourhood and to consolidate their shelter to become part of 'shanty towns' or slums. The third category is 'crisis homelessness' caused by a personal or household crisis such as divorce, family breakup or bereavement. People in this category have little choice about their situation and usually receive little help from government or voluntary organisations. Examples of the first two categories are provided in Case Study 10.1 of homelessness in China.

Case Study 10.1   China

With the opening up of the market economy, and the lifting of some restrictions on moving around the country, came an increase in what is known as the 'floating' population, those people who have left their original place of residence, where their household registration or *Hukou* is located. The majority of housing in China is still allocated by the employer and associated with the workplace. However, because jobs in state and collective industries and institutions are only available to local residents with local *Hukou*, the 'floating population' are not entitled to them, and thus, they are limited in their access to workplace allocated housing. As they cannot afford to rent or buy on the private housing market they are restricted to 'aggregated villages', designed especially for them.

Many of these Chinese 'floating' people fall into the category of supplementary homelessness as they form few relationships with local people and maintain close links to their areas of origin. These links manifest themselves in frequent home visits, possession of farmland at home, remittance of money to the original household and an intention to return to their place of origin. Ironically, although unemployment is a major cause of homelessness in China, the unemployment rate amongst the floating population in urban areas is lower than amongst the local urban population.

Whilst the most obvious driving force behind survival homelessness is household economic survival, there is another form of survival which drives people to accept, even choose, situations which would be considered as homelessness in the West, that is unit or household survival. Migrant labourers in China were considered to be predominantly in the 'supplementary homeless' category, because they developed few ties and networks in the urban locations and generally returned to their place of origin once the homelessness had served its purpose in improving their financial situation. However, other homeless Chinese families fall into the survival category as they are homeless for entirely different reasons.

Because of the strict 'one family one child' policy in China, some couples who wish to have more children choose to leave their household registered place. The women in these 'over-procreated' families are all regarded as Sanwurenyuan, meaning, without official identification card, because they cannot get the necessary official Temporary Living Permit for their new location without the Family Planning Certificate granted by their native neighbourhood. Their homelessness is not as a means of economic survival but rather it is linked to a desire to survive as the type of family they choose to be.

Source: Based on Speak (2004).

The term 'homelessness' of course means the lack of a home. Home can mean many things as we showed in Chapter 5, and it is rare that all of these are included in any official definition of homelessness. Nevertheless, the inclusion of the word 'home' does seem to indicate that it is not only physical elements, but also emotional ones that are usually included in the term, even if the range of factors taken into account varies considerably between countries.

Jacobs et al. (1999) identified what they termed the minimalist and maximalist discourses of homelessness that vary in their definitions of the causes of homelessness and, therefore, of the appropriate means of dealing with it. The minimalist discourse tends to see homelessness as a personal failing of the individuals concerned. The maximalist discourse focuses on the structural issues that create the context for homelessness. Some of these will be related specifically to housing circumstances such as a shortage of housing or affordability problems, but others may relate to wider issues such as the extent of unemployment or lack of social security payments, or social care for vulnerable people. In practice, the two discourses are entwined in much public policy as the example of homelessness policy in England shows (see Case Study 10.2). In general, anyone not having a home in England (defined as including both shelter and elements of insecurity) is defined as homeless, but help is only provided to some people and if they meet certain conditions. For example, if someone is defined as intentionally homeless, they are not deemed to be deserving of help. Intentionality can include the act of not paying rent or mortgage repayments because of a low income. Thus, categories of deserving and undeserving are preserved as they have been in much social policy in many countries. The rationale for this is first to ration help when resources are perceived to be short and second, preserving incentives to work.

Case Study 10.2 Homelessness in the UK: A rights-based approach

The Housing (Homeless Persons) Act 1977 was passed well after the heyday of the UK 'post-war consensus' in welfare, and in fact came into force in 1978, just before the election, in 1979, of the first of the radical Conservative Governments under Margaret Thatcher (for a history of its enactment, see Somerville, 1999). This legislation was notable, amongst other things, for being posited on a fundamentally 'structuralist' analysis of homelessness, as driven primarily by housing market failure. The original Act covered all of Great Britain, but was subsequently incorporated into separate legislation for England/Wales and Scotland, and was extended to Northern Ireland in 1988. There has been significant divergence in this statutory framework across the UK since the mid-1990s (Fitzpatrick et al., 2009), and so the discussion throughout the remainder of this paper is restricted to England, where the relevant legislation is now contained in Part 7 of the Housing Act 1996, as amended by the Homelessness Act 2002.

The statutory homelessness system in England

Strictly speaking, the 'main homelessness duty' of local authorities in England is to provide temporary accommodation until 'settled' housing becomes available, found either by the household itself or by the local authority. In practice, this settled housing is almost always secured by the local authority that owes a duty under the Act, and in the great majority of cases the duty is discharged via the offer of a social rented tenancy. By international standards, the definition of homelessness used in this legislation is remarkably wide (Fitzpatrick et al., 2009), as it includes not only those who are 'roofless', but also households which cannot be 'reasonably expected' to live in their current accommodation. However, in order to be owed the main homelessness duty homeless households must also be 'eligible' for assistance (asylum seekers are the main ineligible group), 'unintentionally homeless' (i.e. have not brought about their homelessness through their own actions or inaction), and in 'priority need' (the principal priority need groups are households which contain dependent children, a pregnant woman or a 'vulnerable' adult). If a household meets all of these criteria but has no 'local connection' with the authority to which they have applied, the main duty can be transferred to another UK authority with which they do have such a connection (unless they are at risk of violence in that other area). Homeless applicants can challenge local authorities' decisions under this legislation by way of judicial review, and a statutory appeal to a relevant court (on a point of law) is also now provided.

From its very inception, the legitimacy of the homelessness legislation has been questioned. It was memorably described in the original Parliamentary debates as a 'charter for scrimshankers and scroungers'. More soberly, it has been argued that the 1977 arrangements are unfair because they grant homeless households priority in social housing allocations over others with 'comparable underlying housing needs', meaning non-homeless households which have suffered similar or greater levels of 'long-term housing deprivation' (Fitzpatrick and Stephens, 1999).

Two reviews of the 1977 Act under Conservative Governments did not result in action, but following the 'Back to Basics' 1993 Conservative Party Conference the Government yet again turned to reform of the legislation. The 'moral hazard' objection to the 1977 Act was foregrounded, with the government arguing that it created a 'perverse incentive for people to have themselves accepted by a local authority as homeless' (DoE, 1994, p. 4). Many commentators have highlighted the lack of empirical evidence for such conscious manipulation of the homelessness system (Cloke et al., 2000), but there has hitherto been little serious attempt to investigate its existence, and the scope for collusion is clearly significant given that family and friend 'exclusions' is a key reason for acceptance as homeless (Fitzpatrick and Pawson, 2007). Likewise, while the idea of an incentive to 'go homeless' has been called into question by the poor quality of accommodation allocated to homeless households under 'one offer only' policies (Somerville, 1999), in high-demand areas (such as London) it is conceivable that households may 'manufacture' a homeless status if they know that this is their only realistic option for being rehoused.

Whatever the merits of their case for reform, the Conservative Government's Housing Act 1996 was duly passed and reduced local authorities' maximum duty towards homeless households to securing temporary accommodation for a time-limited period of two years. The separate 'homeless route' into social housing was also eliminated through the introduction of the 'single housing register', and homeless households were removed from the list of groups for whom a 'reasonable preference' had to be given in council house allocations. However, as predicted (Fitzpatrick and Stephens, 1999; Somerville, 1999), the 1996 Act had a relatively limited impact in practice, as only the minority of local authorities unsympathetic to the legislation acted on this weakening of their rehousing duties towards homeless households (Cloke et al., 2000). Thus the 1977 framework emerged relatively unscathed from 13 years of hostile Conservative Government, albeit that other Conservative housing policies (notably the Right-to-Buy for council tenants combined with a virtual cessation in council house building) seriously undermined some local authorities' capacity to respond to 'statutory demand' (Malpass, 2005).

The Labour administration that came to power in 1997 largely reversed the 1996 Act via the Homelessness Act 2002 and secondary legislation, and at the same time expanded (modestly) the priority need groups and introduced a new duty on local authorities to prepare homelessness strategies. However, the context for the operation of the homelessness legislation became increasingly difficult during Labour's time in office. A prolonged period of rising house prices (before the credit crunch in 2007 and the subsequent recession), combined with a sharp drop in the supply of social sector lettings, increased access pressures for low-income households, notwithstanding some expansion in the limited supply of private rented dwellings available to this group (Rugg and Rhodes, 2008).

These growing housing pressures meant that the number of statutorily homeless households in temporary accommodation started to rise, peaking at 101 300 in December 2004. The Labour Government then introduced an official target to halve the number of households in temporary accommodation, from this December 2004 figure, by the end of 2010. Critical to pursuing this target was the establishment of a preventative 'housing options' approach, strongly promoted by central government (Pawson et al., 2007). Households approaching a local authority for assistance with housing are now usually given a formal interview offering advice on all of their 'housing options', which may include services such as rent deposit guarantee provision, debt advice or family mediation designed to prevent the need to make a statutory homelessness application. The implementation of this prevention agenda has been associated with an extraordinarily sharp drop in statutory homelessness acceptances (down by around 70 per cent since 2004), and the temporary accommodation target was met ahead of schedule. Questions have been asked about the extent to which these trends represent 'real' reductions in the level of homelessness, or arise, at least in part, from increased local authority 'gatekeeping' which is (informally) raising the 'threshold' for statutory assessments (Pawson, 2007).

One might expect the Conservative-Liberal Democrat Coalition Government elected in the UK in May 2010 to be hostile to the homelessness legislation, certainly within its (dominant) Conservative quarters. At the time of writing, the Coalition Government had issued a housing reform consultation paper wherein the homelessness legislation was proposed to remain ostensibly intact, but the 'offer' to homeless people was to be significantly altered by introducing 'compulsory' (i.e. without the applicant's consent) discharge of duty into fixed-term private sector tenancies (Communities and Local Government, 2010). Moreover, Coalition plans to 'localise' the access rules for local authority waiting lists, and to give social landlords discretion to grant 'fixed-term' rather than 'lifetime' tenancies to new tenants, speaks to an agenda radically at odds with the 'national minimum standard' ethos underpinning the statutory homelessness framework. Throughout its history criticisms of the statutory homelessness system have come not only from those who oppose it in principle, but also from those who fundamentally support it, but regret its apparent operational deficiencies. First, concerns have focused on the social, health, educational and other impacts of prolonged stays in temporary accommodation on children in homeless families (Thomas and Niner, 1989), with parents in these families describing a persistent feeling of life being 'on hold' (Sawtell, 2002). Second, inappropriate types of temporary accommodation have long since been a source of complaint from homelessness pressure groups, with specific concerns about the quality and suitability of Bed & Breakfast (B&B) hotels (Carter, 1995; Niner, 1989) leading eventually to the prohibition of the long-term use of this form of temporary provision for families with children. Third, it has been suggested that multiple moves between temporary accommodation addresses can be extremely disruptive for homeless families (Homelessness Directorate, 2003). There are also long-standing concerns about the quality of the settled housing that homeless households access at the end of this statutory process (Somerville, 1999).

The coalition years, austerity and welfare reform

The Conservative-led UK Coalition Government entered office in May 2010 on a platform of austerity and a linked radical welfare and housing reform agenda. The Localism Act 2011 brought about some changes that made it easier for local authorities to discharge their main rehousing duty need via fixed-term private tenancies, but in practice more significant was the Coalition agenda to drastically reduce welfare spending, particularly on Housing Benefit. All enumerated forms of homelessness have subsequently escalated in England, including the most extreme form, rough sleeping, wherein the official numbers have doubled since 2010 (see Figure 10.1) (Fitzpatrick et al., 2016). While the official rough sleeper statistics in England have some very well-documented methodological weaknesses, there is no mistaking this national upward trend. More robust data for London also shows an accelerating upward trend, particularly amongst Central and Eastern European migrants, but also UK nationals. Bear in mind that most of those affected by rough sleeping will be single people, generally assessed as lacking 'priority need' in England, and as therefore having no rehousing entitlements under the homelessness legislation.

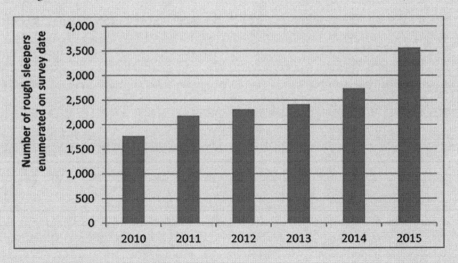

Figure 10.1 Rough sleeping in England, 2010–2015

> Rising homelessness in England is, we would argue, the result of deliberate policy choices rather than the post-2008 recession: as noted above, previous recessions have not necessarily witnessed a rise in homelessness. International comparative research also indicates that it is the welfare safety net, especially housing allowances, that usually operate to 'break the link' between losing a job and homelessness in developed welfare states. So, it is the weakening of that safety net that has driven up homelessness (Stephens et al., 2010).
>
> Source: Based on Fitzpatrick and Pleace (2012) and Fitzpatrick et al. (2016).

The integration of personal and contextual factors in the causation of homelessness is a key facet of recent research on the topic. Neale (1997) pointed to the lack of integration, with academic and theoretical approaches mirroring the personal and contextual split that featured in political discourse. However, more recently there have been a number of contributions that have sought to overcome the divide. Clapham (2003) has used the concept of a pathway to examine the opportunities and choices that people have and the factors that impact on this. This has led to investigation of pathways into, through and out of homelessness, examining the interaction between personal and contextual factors. The pathways approach places emphasis on both the personal and structural elements in the causation of homelessness and their interaction. At the same time, it focuses attention on the holistic nature of the experience that has causes and implications for many phases and aspects of life. The approach also highlights the temporal and spatial aspects of homelessness. It is misleading to see homelessness as a static phenomenon, as its causes evolve over time and the experience is often not a one-off event, rather it is a process that can be extended over a considerable period of time.

In a critical realist approach, Fitzpatrick (2005) has identified the personal and contextual factors that increase the risk of homelessness and the direct factors that trigger an episode. The most important trigger factors vary between countries, but the more common are eviction from a rented property, relationship breakdown and loss of employment. The underlying causes of homelessness also vary between countries and depend to some extent on the housing regime. In some countries, emphasis is placed on personal factors such as drug abuse and mental health problems, whereas in others there is a focus on the inadequacies of housing and social policies. Bramley and Fitzpatrick (2018) follow Shinn (2007) in arguing that countries with well-functioning housing and labour markets, relatively low levels of poverty and inequality and generous social security policies will have a relatively low level of homelessness and homeless people will tend to have complex personal problems. She argues that the reverse will also be true. However, there is no definition here of what a well-functioning housing market entails. Toro (2007) shows that 'lifetime homelessness' was significantly more common in countries such as the UK and USA with high levels of inequality than in Belgium, Germany and Italy. Shinn (2007, p. 658) concludes 'The association between the flatter income distributions and higher social welfare expenditures in continental Europe than in the United States and the United Kingdom and the lower rates of homelessness is suggestive, although not evidence of a causal link'. European Commission-funded comparative research also supports the proposition that 'welfare regimes' impact profoundly on the scale, causes and nature of homelessness (Stephens et al., 2010). It further suggested that housing market conditions may have a more direct effect on homelessness than labour market change. As Bramley and Fitzpatrick (2018, p. 114) conclude

> [I]t has made clear that, in the UK at least, homelessness is *not* randomly distributed across the population, but rather the odds of experiencing it are systematically structured around a set of identifiable individual, social and structural factors, most of which, it should be emphasized, are outside the control of those directly affected.

The extent and type of homelessness can be seen as an indicator of the success of housing and social policies in general. For example, Case Study 10.2 of homelessness in England shows the almost doubling of rough sleeping since 2010, in contrast to its decline over the previous decade, to historically low levels at that time. It seems unlikely that the number of people with personal social problems will have varied so much or so systematically, so the variation is most likely explained by changes in housing and other public policies, or to national phenomena such as the state of the economy and labour market changes. Nevertheless, it is difficult to draw direct inferences on the success of the housing regime from the extent of homelessness because of the many other policy, personal and structural factors that are involved. Case Study 10.3 shows the housing policy factors that lead to homelessness in Argentina and the particular form that homelessness takes in this situation.

Case Study 10.3 Homelessness in Argentina

Despite the city constitution's declaration of housing as a universal right and the creation of programs that formally recognize a municipal "housing crisis", little has been done to resolve the housing deficit in Buenos Aires. For many struggling to remain inside Buenos Aires, living in informal hotels or squatting in once-empty buildings located throughout the city offers some security and benefits not available outside of the city or in its informal settlements (*villas miserias*). Yet, these housing options are highly precarious, with most residents eventually facing eviction and displacement.

As a housing strategy of poor residents who are unable to access formal housing, informal hotels are often state-sanctioned run-down family hotels where families live in one bedroom and share a kitchen and bathroom with other families, all the while paying rent to a manager who also lives on the premises. Similarly, *casas tomadas* are usually empty buildings or apartments that have been taken over by individuals who then rent out or sell rooms to make a profit. In many cases, these two forms of housing are indistinguishable, especially for the residents, who are often surprised when they receive a notice of eviction from the owner or the city government.

> Inside, each family lives in one bedroom where they usually watch television, prepare food, eat, study, and sleep, and store all of their belongings. Nuclear or extended families of up to four or five may live in a room measuring approximately 12 × 12 feet or smaller, sometimes without windows or ventilation. Shared kitchens may have one or two ovens or small gas burners. Although not necessarily clean, they are usually orderly because residents must keep all of their belongings in their bedroom, or they run the risk of having them stolen. Bathrooms are rundown, sometimes with dripping sinks and showers, broken mirrors, overflowing trash bins, and toilets that constantly leak. Some of the buildings do not have hot water, and electric and gas connections are often old and haphazardly installed.
>
> Since 2000, the city government has attempted to deal with the issue of evictions and the housing crisis through programs that temporarily address the immediate threat of eviction. The *Programa de Atención para Familias en Situación de Calle* – the subsidy – is a program sponsored by the city government that offers a monthly quota to families evicted from their home. On the day of eviction, families must go to the Welfare Office (*Desarrollo Social*) to receive the first quota. Afterwards, they can receive a monthly allotment for up to twelve months. In order to be eligible, families must be in the house on the day of eviction. Later, they must follow a series of procedures to continue to receive the subsidy. These procedures and requirements involve a substantial amount of time, travel and money, often without any guarantee they will receive the subsidy each month (Auyero, 2012).
>
> Source: Based on Munoz (2017).

The housing policies that are associated with the extent and experience of homelessness are those that impact most on low-income people as they are the ones for whom misfortune more likely includes the lack of shelter. The easier the access of low-income people to good quality, usually rented, housing, the less likely they are to be homeless. State-provided housing usually fulfils this function, and the example of Sweden where municipal housing companies have recently been expected to behave in a market-oriented manner shows the difficulties that may occur when the traditional social role of public provided housing is not fulfilled (see Case Study 9.5). In some circumstances, private renting may also provide accommodation for low-income people, although the barriers to access can be greater in this sector. Access to the tenure may involve the need for a deposit and the existence in particular localities of landlords willing to let to benefit claimants. In general, the more effectively the housing regime caters for low-income households, the less likely is there to be homelessness. At the point of homelessness, access may have to be facilitated by public or voluntary organisations prepared to negotiate access or provide security for the deposit, so a different set of factors may come into play. The experience of low-income households in their housing may also be important, including factors such as security of tenure, length of contracts and permissible reasons for eviction.

Affordability issues may also be an important influence on the extent of homelessness through factors such as the extent to which housing allowance systems meet the full costs of renting, as well as the level of rents. Other public policies that may have an influence on the level of homelessness include general welfare benefit levels and conditions. As well as the overall level of benefits, any delays or breaks in payments, such as have occurred in England because of welfare sanctions and the introduction of the Universal Credit system, may cause difficulty with meeting housing costs whilst maintaining spending on other essentials. Social and health services may also influence homelessness through helping people overcome disabilities and personal issues that enable them to deal with life problems such as their housing situation. When homelessness occurs, health and social services may be important in helping people to cope with their situation. Therefore, reductions in such services, as has occurred in England through the general reductions in public expenditure under 'austerity' have been partially responsible for the increase in rough sleeping and make solutions to homelessness, when it occurs, more difficult.

Certain personal characteristics are associated with homeless people. For example, they are more likely to have poor physical and mental health as well as problems of drug or alcohol abuse than the general population. The causality of these factors is open to doubt though as they may have caused homelessness or been caused or exacerbated by the experience of homelessness. Trigger factors for homelessness can include divorce and arguments or violence at home.

Homelessness is a complex phenomenon with many causes and symptoms, and every homeless individual has a unique homeless pathway. Individual circumstances can change quickly as homeless people pursue a route through the complex web of opportunities and constraints that face them. Many homeless pathways involve constant changes of abode as people move rapidly between rough sleeping, sofa surfing and hostels. The complexity and changing nature of homelessness make it a difficult area for government intervention.

Objectives of government intervention

The impact of homelessness on homeless people themselves and the wider population puts pressure on governments to intervene to prevent homelessness and deal with it when it occurs. It is widely accepted that homelessness excludes people from the wider society by cutting them off from employment and makes it more difficult for them to access public services such as health care. The impact on their mental and physical health has already been alluded to. Homelessness also has an impact on society in general because of its visibility and impact on the daily lives of people in towns and cities. But intervention may have very different objectives. One may be to deal with the visible aspects of homelessness by eliminating rough sleeping, although this may be undertaken in a way that does not improve the lives of those sleeping rough. Examples may be the introduction of design or streetscape features such as spikes as deterrents where rough sleeping takes place, or measures to prohibit rough sleeping in certain locations or to ban services that are provided on the streets such as meals or to deal with activities that can be associated with rough sleeping such as begging. In some countries where rough sleeping becomes established in certain locations, there can be measures to destroy temporary shelters or to forcibly move people on. These measures may not improve the lives of homeless people or solve the homelessness problem as it may resurface in other ways and in other locations.

Other mechanisms may be focused on rough sleeping, but take a more long-term approach based on the needs of the rough sleepers themselves. Therefore, the aim of government intervention may be to improve the lives of people who have become homeless by helping them to achieve secure and permanent housing. Initiatives here usually involve support for the homeless person, although as we shall see, there are many ways of achieving this.

Help may not be solely focused on those who have become homeless but may be aimed at preventing homelessness. Measures to achieve this may range from advice for those most at risk to general issues to help low-income households, as we have discussed in previous chapters. With this aim, the eradication and prevention of homelessness becomes one aspect of a holistic housing policy rather than a stand-alone element. This opens up the link between homelessness policy and the general housing regime. As argued earlier, housing regimes that put the reduction of inequality and the situation of the poorest at the heart of their strategy are more likely to be characterised as having low rates of homelessness.

Tools and mechanisms

The complexity of homelessness is reflected in the strategies and policies pursued by governments to deal with it. These interventions may involve a number of features that we identify here before examining some specific policies and programmes that have included some of these features in a particular combination. Here we do not consider some of the broader factors and policies that could influence the incidence and character of homelessness such as the extent of income and wealth inequality, labour market policy and social security provision as to do so would be outside the scope of the present volume. However, these factors could be important structural issues that have a wide influence on the society including on homelessness. As we have seen in previous chapters, housing problems can often be the symptoms of wider issues. However, here we focus on the housing policies and programmes aimed to deal with homelessness.

Housing regime features

What is included here are elements of general housing policy that could influence the incidence of homelessness. Examples could include the supply and affordability factors and policies examined in previous chapters. Clearly, both of these factors could impact substantially on the housing opportunities of low-income households and so have an impact on the likelihood of homelessness. The evidence was given earlier of a relationship between the type of housing regime adopted and the extent and nature of homelessness. A neoliberal regime with substantial inequality and little intervention to aid low-income households is likely to feature substantial levels of homelessness.

Also, the nature of homelessness and the appropriate government response may be influenced by contextual factors in the housing regime. For example, Sweden adopted a view of homelessness as an individual culpability with the responsibility for dealing with it given to local social services authorities, who worked with individuals to attempt to deal with their particular problems (Clapham, 2017). This response may be appropriate in a situation where there is no shortage of appropriate and affordable housing, and this situation was achieved in large measure in Sweden in the 1980s after the million programme. However, after the economic problems of the 1990s and the reduction or withdrawal of many housing subsidies, this favourable situation no longer exists, and there are major problems of shortage and affordability, especially in areas of high housing demand such as central Stockholm. As a consequence of this and a relatively high rate of immigration, homelessness has increased, and the system, devised in very different circumstances, is under considerable stress. Rough sleeping is now present and visible on the streets of many Swedish cities (Clapham, 2017).

The main point of this section is that homelessness cannot be considered in isolation from housing policies in general and that the main strategy to minimise homelessness and to deal with it when it occurs may be through general housing policies, and particularly those that impact most strongly on low-income people.

Housing rights

One way of dealing with homelessness is through the provision of legally enforceable rights. It was mentioned earlier that housing could be considered to be a fundamental human right as recognised by the United Nations. If this is the case, then homelessness is a breach of those rights that demands government action to rectify. Therefore, one approach to homelessness is to create a framework of individual legal rights that a homeless person (or one in danger of homelessness) can use to get appropriate help from a relevant government agency.

In England, there is a right for homeless people (over the age of 18) to receive settled housing if they are homeless, but they have to pass three tests: they have to be unintentionally homeless; they must have a local connection to the area where they are seeking help; and they must be 'in priority need' and vulnerable (which can cover people deemed to be vulnerable such as women with children, young people and people with mental health problems or disabilities). 'Priority need' has to be assessed on an individual basis, but in general it seems to exclude single people without children and particularly young men. This has led to a distinction between what has been termed 'official' homelessness i.e. that which is recognised officially and which brings a right to housing and 'unofficial' homelessness that covers those who do not receive the right, although they may receive help and advice in finding accommodation that may involve being given a list of properties available in the private rented sector. The nature of the settled housing that homeless people have a right to has been defined differently over the years and has been watered down in the last few years (see Case Study 10.2).

Scotland has gone further than England and has abolished the intentionality and vulnerability clauses, therefore opening the route to housing for all with a local connection. This still excludes new immigrants and rough sleeping by immigrants is still visible in major Scottish cities. In some countries, such as Sweden, there is a right to shelter rather than a right to housing. In other words, local authorities have a duty to provide emergency shelter for those who need it, although of course we have remarked above on the difficulties in making a distinction between shelter and housing. In Sweden, the duty has been one of the factors supporting the 'staircase' model and the accommodation of homeless people in supported accommodation rather than in mainstream housing.

These examples fall short of providing a right to housing as the rights given are circumscribed, and the final outcome can be unsatisfactory given the definition of permanent housing, but they show one approach to dealing with homelessness through a rights-based approach.

Shelter provision

Many definitions of homelessness include the element of shelter. This is not always the case as we showed earlier, as someone may be defined as homeless if they have insecure or temporary housing even though they may have a physical shelter. Nevertheless, the provision of shelter is one of the key elements of government provision to deal with homelessness. A key question is the nature and quality of that shelter. It may take the form of permanent housing or may be some form of supported accommodation such as a hostel, sheltered housing or group home (for a review of possible forms, see Clapham, 2017). Hostels or similar forms of accommodation such as emergency shelters may be made available in the short term in an emergency where no alternative accommodation is available. Alternatives are bed and breakfast accommodation.

This form of accommodation can be regarded as 'intermediate' in that it is seen as providing shelter and possibly respite and an opportunity for an assessment process (see next section) but not a permanent home. Therefore, where it is used, there is usually an issue about move-on to more appropriate long-term accommodation. This move may be after a set period of time or may be subject to an assessment process or may be conditional on observance of certain behavioural codes (such as abstinence from drugs or alcohol). There may be a series of different forms of accommodation with people moving from one to another as their circumstances change or they are judged to be able to cope with or to deserve the next stage. These approaches have been labelled the staircase model which will be considered in more detail later in the chapter.

Personal support

One element of help for homeless people may be the provision of social support to enable them to cope with any problems that are associated with their position and may have contributed to their homelessness or be harming their ability to escape the situation and to be able to maintain independent living. The categories of support provided may vary considerably. One category is practical help in everyday tasks of living independently such as cooking, paying bills, managing money and so on. Another is dealing with specific health or social problems such as mental health issues or alcohol or drug dependency. A further category may be related to skills needed in the workplace to enable people to generate sufficient income to be able to acquire and sustain permanent housing. Personal support may be delivered in supported accommodation such as a hostel or may be 'floating' in the sense that it is available in any location. The support may be personally tailored to suit the needs of the individual or it may be generic.

Prevention

Much government intervention in homelessness is aimed at dealing with the situation once it has occurred, but there is an increasing focus in some countries (such as the UK; see Case Study 10.2) on trying to prevent it occurring. This can be done in a number of ways. For example, support can be given to households to enable them to cope in their present housing and overcome problems that may lead to them losing their home, such as difficulties in paying their rent or in mortgage repayments. Advice can be given on benefits available or help with budgeting. Another example is action in mediating between the household and other parties such as landlords or mortgage providers to forestall homelessness. Information can be given to households about housing opportunities and advice on the best way to access them.

Housing First and the staircase model

The two main forms of government intervention to deal with homelessness are labelled Housing First and the 'staircase' model. The Housing First approach grew out of the perceived failings of the latter, and so we will start with this.

There are many versions of the staircase model (as there are of the Housing First approach), but the essence of the approach is the use of a spectrum of different accommodation and support options for homeless people that they move through, depending on their own abilities and needs. An example would start with emergency accommodation in a shelter for those newly homeless. After an assessment, a homeless person may move on to a hostel or into a group home with other people with similar abilities and needs. Both forms of accommodation will usually have support that is given to the homeless person, although this may vary considerably in form and extent. When a resident is perceived to be ready, they may be offered a place in an independent apartment with the lease being held by the municipal authority who may still provide support services. In this situation, the homeless person does not have their own tenancy and so has fewer rights than a tenant and may be subject to special conditions on their behaviour. The final prize is a full tenancy when people are judged to be able to cope and to deserve it. There are many variations on this basic model in different countries. For example, in Britain there is a shortage of emergency and hostel accommodation, and many people accepted as homeless (often a young woman with children) will be housed in bread and breakfast accommodation usually consisting of one room (sometimes with shared bathroom facilities). In areas of high housing demand this situation may persist for a considerable time before more secure and appropriate accommodation becomes available.

The drawbacks of the staircase model are legion (see Sahlin, 2005; Johnsen and Teixeira, 2010). First it does not often seem to work on its own terms. People are supposed to enter at the bottom of the staircase and leave at the top, but the evidence is that few leave. Rather they become 'stuck' on one of the stages and never reach the goal of an independent tenancy. Second, and directly flowing from the first, it is an expensive form of provision that demands specialised facilities and extensive support, and with few people progressing through the system, the financial commitment is continuous. Third, the model empowers professionals and providers rather than the homeless people themselves who have to 'prove' themselves in order to move on to the next stage and to convince professionals that they are ready. It is perhaps not surprising that few homeless people gain the confidence and self-esteem to climb to the top of the staircase. Fourth, although social support is offered as part of the model, its form has often been challenged as it is seen as a 'holding operation' rather than as a resource to foster onward progression. This may be due to the shortage of resources, but also to the ethos of the forms of provision. For all these reasons, provision in the staircase model tends to become 'silted up' with little movement and so few opportunities at the bottom.

These deficiencies were recognised by advocates of the Housing First approach that was put forward to overcome them (for example, see Tsemberis, 2010). The approach has three basic elements: the first is a person-centred approach in which the homeless person is in control; the second is the initial priority given to the provision of permanent, independent housing in the belief that once an appropriate home has been secured, people are better able and prepared to deal with any underlying issues; third is the availability of personalised support. In more detail, the approach is based on seven fundamental principles (see Johnsen and Teixeira, 2010). The first is the immediate provision of independent accommodation in mainstream housing. In many cases this has been in private or state rented apartments leased by the Housing First provider. Second, there is no requirement of what has been termed 'housing readiness', that is a professional assessment of the ability to be able to cope with independent living that is at the heart of the staircase approach. Third, there is no requirement that people refrain from drug or alcohol behaviours as a condition of help nor that they accept treatment for these or other conditions, which again is a contrast with many versions of the staircase model. Fourth, the provision of permanent housing and support that is not conditional or temporary. Fifth, the provision of integrated and comprehensive community-based support. Sixth, is a person-centred approach that empowers the individual homeless person and gives them opportunities and choices in order to achieve their own goals that may differ from those of others. It is argued that this empowerment creates the self-esteem, confidence and motivation that result in the person taking control of their own life and using the support to deal with their problems.

Finally, programmes should target the most vulnerable homeless people, that is those that face multiple obstacles to housing stability.

The Housing First concept has been adopted in many countries and, as with the staircase model, there are many variants of the approach, some of which bear little resemblance to the original idea. Case Study 10.4 gives an example of the approach in Australia where small-scale Housing First projects have been implemented in some cities.

Case Study 10.4   Housing First in Australia

Johnson et al. (2012, p. 5, citing Tsemberis, 2010) characterise the Housing First approach as having five key elements:

1   Rapid access to permanent housing: Housing First approaches usually emphasise the importance of offering permanent housing with support, as opposed to transitional or emergency accommodation or mainstream tenancy with no support. Housing is dispersed, although there are some congregate models such as 'Common Ground' with support services nearby or even on site.
2   Consumer choice: Persons actively choose to participate. There is no conditionality attached to housing (in relation to complying with treatment or engagement with support services).
3   Separate support services: Support services are provided by different organisations to those who provide the housing. Support services are also individually focussed. Support is provided in a holistic manner – casework, clinical, legal, specialist counselling, and other services based on individuals' needs. Importantly support is not time-limited. Rather, it is assumed that support will need to be ongoing or at least available in the long-term.
4   Recovery as an ongoing process: Housing First approaches to homelessness understand housing to be the first step in recovery. Housing is not conditional upon engagement with treatment and support services and there is a recognition that recovery takes time.
5   Reintegration into the community: The final key feature is the integration of Housing First housing into the community. Housing First housing should be located in 'normal' neighbourhoods, rather than being separated. Central to the 'reintegration into the community' is the ancillary goal of Housing First programs of moving people into employment or education.

> Although there has been widespread articulation of the Housing First approach in homelessness strategies, from the National, State and local government level, and despite the written policy commitment, the implementation of the Housing First approach has been partial. In most cases it has been small-scale, 'experimental' or pilot projects, some of which were not provided with adequate access to designated housing stock (Phillips et al., 2011). These programs have relied on the existing social housing stock rather than 'new build' housing and/or headleasing. Access to suitable housing is regularly reported as a key barrier to success in evaluations and reports (Johnson and Chamberlain, 2013).
>
> Source: Based on Blunden and Drake (2016).

Many governments have been swayed by the argument that the Housing First approach is cheaper and more effective, but have not been able to implement all the elements that have made the approach viable and successful (see the example of Australia in Case Study 10.4 above). This is partly because the full implementation demands circumstances that are not universal. The first requirement is available permanent accommodation for homeless people to be moved into. In many countries, this is not available anywhere, but even where the national situation is appropriate there may be many locations, particularly in the major cities with high housing demand, where there is a shortage of affordable housing that has contributed to the existence of homelessness. Therefore, to find accommodation for Housing First schemes, affordable housing has to be diverted from others, and this could be politically difficult as well as only redistributing the incidence of homelessness rather than reducing it overall. The standard and location of the accommodation is important because it has to have the desired impact on feelings of security and self-esteem if homeless people are to have the confidence and motivation to deal with any underlying problems they may have.

Although the philosophy of Housing First is widely accepted, the difficulties in meeting the extensive requirements have meant that implementation has varied, with some countries taking just part of the approach and others supporting pilot or demonstration projects in specific locations and often on a small scale (see example of Sweden in Clapham, 2017). Nevertheless, the results from evaluations of Housing First initiatives seem to be very positive as we shall see in the next section.

Evaluating homelessness policies

The complexity of homelessness makes any evaluation of its impact difficult. However, a number of attempts have been made to evaluate specific aspects of the outcomes (for reviews, see Johnsen and Teixeira, 2010; Woodhall-Melnik and Dunn, 2016). For example, evaluations of Housing First have focused on the cost to public agencies compared to alternative forms of provision (for example, see Larimer et al., 2009) where Housing First was found to provide cost savings in the US context, although this does not necessarily mean that the same would occur in other contexts where the provision and funding of health and care services may differ. Studies have also focused on the outcome in terms of the number of formerly homeless people who succeed in gaining and maintaining an independent tenancy (see, for example, Tsemberis et al., 2004). Here Housing First seems to have been very successful in enabling homeless people to obtain and retain independent living (Woodhall-Melnik and Dunn, 2016). Also, some studies have examined the health outcomes of Housing First with mixed results (for reviews see Johnsen and Teixeira, 2010; Woodhall-Melnik and Dunn, 2016). Although findings have generally been positive, there is some doubt over whether the schemes have been focused on those with the severest problems. As outlined earlier, it is difficult to put together the evaluations of different Housing First programmes, because they vary considerably in their form and content, despite sharing the same basic philosophy. Also, many schemes have only been in existence for a short while and so long-term results are not available. There is also a shortage of studies that examine subjective well-being of participants in the schemes. Padgett (2007, p. 1934), for example, reported that, while Housing First offered homeless people ontological security, that is 'a sense of wellbeing arising from constancy in one's social and material environment', other core elements of well-being such as 'hope for the future, having a job, enjoying the company and support of others, and being involved in society' had only been partially achieved. Housing First seems to be an effective housing policy tool, but its impact on the health and overall well-being of homeless people seems to be more limited even though it passes comparison with traditional models in this regard and the short timeframe of evaluations may explain the relative lack of progress in what may be difficult and complex conditions.

Despite these drawbacks, the results from the schemes seem to be very positive.

Housing First seems to be lower cost and result in better outcomes for homeless people than the traditional alternatives such as the 'staircase model'. This seems enough evidence to show that the basic philosophy of the approach is sound and offers a very positive way forward. The problem is that effective implementation of the approach needs access to an appropriate amount of good quality and affordable housing as well as relevant social support. Therefore, the success of Housing First shows the value of a general housing policy that provides these fundamentals for low-income people at risk of homelessness.

In principle, the rights approach delivers a widespread Housing First approach. However, rights are usually limited to certain individuals and circumstances and have been difficult to operationalise, resulting sometimes in a poor experience even for those who qualify for help. In England, the number of people in temporary accommodation for long periods and the unsatisfactory placement in private rental housing are evidence of this (Crisis and Shelter, 2014). As with Housing First, the rights approach shows the importance of the availability of the fundamentals of housing and support in dealing effectively with homelessness.

Conclusion

Homelessness is the ultimate failure in housing, both on an individual and a societal level. For the individual, it can lead to poor physical and mental health and exclusion from life opportunities such as employment and access to public services. For a society, the extent of homelessness is an indicator of the success or failure of its housing and social policies towards the poorest sections of the population. A large number of factors can impact on the likelihood of homelessness, ranging from individual characteristics and choices to general economic and labour market conditions and the level of inequality in a society. In between are a wide range of public social policies such as social care and health as well as general housing policies and those aimed specifically at homelessness. The wide scope and complexity of these factors makes any simple connection between housing regime and homelessness problematic. Despite this, there is some evidence that neoliberal regimes such as the USA, Australia and the UK have a higher rate of homelessness than other housing regimes. But even within the neoliberal group, it seems that the extent of public health and social care as well as housing provision in the UK when compared with the USA or Australia can have a positive impact on homelessness.

The nature of homelessness can make dealing with it when it occurs problematic. The UK has adopted a rights-based approach to homelessness, with Scotland and Wales taking this approach further than England. The granting of rights to homeless people reflects the principle of housing being a universal right rather than a marketed commodity and offers a way for homeless people to improve their situation. But, the rights are usually limited to a defined group of people and constrained in what they have a right to. To be effective, rights have to be practical and enforceable, and the lack of access to affordable and secure housing constrains the impact that the possession of rights can have as the result can be many years living in overcrowded, unsuitable and temporary housing even for those given help.

The same constraints operate on the 'Housing First' approach and limit its scope. Where it has been applied, the approach has shown the substantial benefits to the individual and the society of giving homeless people the secure and adequate housing that gives them self-respect and dignity and gives them the confidence and support that allows them to deal with the problems they may have. But 'Housing First' has only ever been implemented as relatively small-scale projects rather than comprehensive national programmes. This is partly about political will, but also reflects the need for access to appropriate housing that the approach requires. Therefore, discussion of homelessness echoes the themes of previous chapters that emphasised the need for affordable and adequate housing for low-income households.

Chapter 11

Environmental sustainability

'Sustainability' is a contested term that has come to have a wide variety of meanings. In its most inclusive form it has been used to indicate social, cultural, economic and environmental sustainability, which encapsulates almost all the contents of this book. The strength of such a wide definition is that it emphasises the important links between all these factors. But the weakness is that the focus on environmental sustainability becomes diluted and can get lost in discussions about social cohesion and neighbourhood integration and other topics. Therefore, the aim of this chapter is to focus on environmental sustainability. The importance of the environmental agenda is highlighted by the identification of a new geological era – the Anthropocene – in which human activities start to have a significant and negative global impact on the Earth's ecosystems (Gough, 2017). Climate change is at the forefront of present environmental debates, and housing is often seen as a key element in the problem and possible solutions.

Environmental policies can be divided into those that mitigate climate change and those that help adaptation to it. Housing is an important topic in discussions about sustainability and climate change because it is involved in both these elements. It is heavily involved in mitigation measures as housing is a major contributor to global warming through the energy expended in the construction and use of houses and is a major user of material resources used in the construction process. Also, housing is important in adaptation to climate change as the location of housing has a major impact on the lifestyles of residents and their need to travel to work and to use public and private facilities as well as to see friends and family. Adaptation to climate change may involve changes in location and lifestyle for many households. Therefore, housing needs to be one of the key elements of any government strategy to encourage sustainable development and to help offset and to cope with the serious and urgent problems of global warming and climate change.

The chapter begins with a discussion of the meaning of environmental sustainability and its basis in resource depletion and climate change. The focus then moves onto housing with a discussion of the main reasons for government intervention on this issue. This is followed by a description of the main tools for policy such as the development of low-energy or carbon-neutral housing or the retrofitting of existing housing stock to improve insulation levels and to reduce energy consumption. In evaluating these policy mechanisms, the links between earth systems and human social practices are explored. The attitudes of households and the discourses that influence them have important impacts on the achievement of environmental aims. Also, the contradictions and complementarities are highlighted between the achievement of environmental and other objectives such as the reduction of inequality or homelessness. In addition, the links between environmental objectives and policies and the housing regime are drawn out with a discussion of the implications of the neoliberal regime for sustainability.

Environmental sustainability and climate change

The term 'sustainable development' came into vogue in the 1980s and reflected the concern with achieving economic development goals without compromising ecological systems and environmental sustainability, thought of as the ability of the earth to sustain the ever-expanding needs and demands of humans. Sustainable development is usually defined as 'development that meets the needs of the present without compromising the ability of future generations to meet their own needs' (World Commission on Environment and Development, 1987, p. 23). Sustainable development has been seen to have a wide scope including social, cultural and economic sustainability. Chiu (2004) argues that this wide definition offers the possibility of sustainability being a holistic concept that can lead to a useful discussion of the inter-linkages in housing policies, and we will return to this later. However, as argued earlier, this wide definition can lead to a lack of focus on key issues of the impact of housing on environmental change and resource use. Therefore, the focus in this chapter is on the role of housing in environmental sustainability. Nevertheless, it is necessary to expand this initial focus to include social and cultural elements, because, as Chiu (2004) points out, the implementation of environmental policies will depend on their acceptability to producers and consumers of housing. For example, the finest insulation will contribute little to offsetting global warming if consumers are unwilling to pay for it and if, when installed, they leave the windows open and so dilute its impact.

In an influential contribution, Raworth (2017) uses the diagrammatical device of a doughnut to show the relationship between Earth systems and human well-being. The outer layer of the doughnut is made up of the nine, different critical earth-system processes identified by the Stockholm Resilience Centre. These are: climate change, bio-diversity loss, nitrogen and phosphorous cycles, stratospheric ozone, ocean acidity, global freshwater supplies, agricultural land availability, atmospheric aerosol loading and chemical pollution. The inner level of the doughnut is made up of the social foundations of human well-being identified in the Sustainable Development Goals identified by the UN in 2015 – one of which is housing. The appropriate space for human functioning is the doughnut between these layers where human well-being is achieved without compromising ecological processes. This formulation puts into sharp relief the intergenerational elements of climate change policy as it highlights the impact of decisions and actions at one point of time on people in the future. For example, it is not appropriate to solve the housing problems of people now if it compromises the earth-system processes that impact on the well-being of future generations. This approach also highlights issues of social justice and inequality and the possible conflict with sustainability objectives. For example, redistribution of housing resources through the reduction of inequality in general and in housing may result in more energy and resource use and so compromise sustainability objectives. The possible contradictions between different state objectives in housing will be considered later in the chapter.

In recent years, major concern about environmental sustainability has been focused on the perceived urgent and serious threat from climate change and global warming, and this will be the focus here. The evidence has mounted of the rate of change that is occurring, and the realisation has dawned that this is likely to have an important impact on patterns of life. Global warming has been linked to the reduction in the ozone layer and the levels of substances in the atmosphere that have led to its decline such as methane and, particularly carbon dioxide. Therefore, there has been increasing focus on the processes by which carbon dioxide has been released into the atmosphere in increasing and unparalleled quantities. Part of the concern has been with processes such as deforestation through urban development and agricultural changes that have removed the earth's capacity to absorb carbon dioxide through photosynthesis. However, the primary focus has been on the use and generation of energy. Currently, much energy is generated through the burning of fossil fuels such as coal, oil and gas that lead to carbon dioxide emissions. Therefore, policies have been devised to reduce energy use, move away from carbon-based energy forms and move towards more sustainable sources such as wind and solar power.

In a product, such as a house, the energy use can be divided into the energy embedded in the materials that are used in its construction, energy used in the construction process itself and the energy used during occupation. The last involves not just the energy used inside the home, but in travel between home and the other places that form the locations of a household's lifestyle such as workplaces, schools and so on. Therefore, the elements of housing that impact on energy use are wide-ranging and cover many of the elements discussed in previous chapters in this book.

As argued above, sustainability is not just about climate change and energy use, as it is also concerned with the use of material resources that are limited to varying degrees. Reid and Houston (2012) criticise the focus on a low carbon discourse and concerns about 'Peak Oil' because they argue it can distract from other issues and promote the idea of a technological fix. Construction of a house involves the use of different materials, some of which may be in short supply or involve the use of methods of extraction that are damaging to the environment in some way. The design of houses may encourage or discourage the use of certain materials, such as plastics, and help or hinder their recycling. There has been increasing concern about the pollution caused by plastics in the oceans and the harm they can cause to humans and to wildlife.

The emphasis on carbon dioxide emissions has tended to result in a lack of attention to other environmental sustainability goals such as water use and the recyclability of materials. The use of water is crucial in many countries and both access to clean water and the marshalling of water resources is vital to ensure availability, especially at times in seasonal cycles when water is scarcer. Even in countries with a plentiful supply of rainwater, there are often problems with effective methods of capture, storage and distribution that mean that supply can be problematic.

Material use is also important. Some materials, such as concrete, have large embedded energy and use finite material resources. Timber is a natural material that stores carbon, but its use can involve high transport costs and, unless it is sourced sustainably and replaced, can reduce the extent of photosynthesis and so contribute to global warming. There have been many examples of the use of traditional, local materials in construction that have little embedded energy and, if sourced sustainably, can minimise the ecological footprint. Examples in the global south may be the use of mud and thatch and, in the developed world, lime products and sustainably sourced wood, as well as traditional cob and straw bales for the walls of a house and sheep's wool for the insulation.

The use of more natural materials is suited to processes of house construction that are more local and empowering for the residents. In many cases, materials can be sourced locally, and construction can be small-scale and flexible in timing and form. Therefore, policies such as sites and services provision fit well with this priority as residents can construct their house with the materials they have at hand and can adapt to meet changing resources and materials. However, there could still be an important role for technology in adapting natural materials and using them in new ways. An example may be the use of laminated wood beams to replace steel as structural elements in a building.

The type and design of houses may impact on the ability of households to live sustainable lifestyles. For example, the availability of garden space is a requisite for being able to grow one's own vegetables, unless recourse is made to available public allotment sites or other communal sites. Some designs have attempted to include employment spaces in dwellings in order to reduce the use of energy in travel to work (see Bedzed example in Jones, 2012).

The existing extent of climate change has created a situation where the incidence of 'severe weather events' has increased substantially, and the accelerating speed of change means that such episodes are predicted to become more frequent in the future. The specific nature of these events may vary according to geographical position. In some countries, it will be the increased risk of flooding that makes residential locations previously inhabited for centuries difficult to sustain. Despite this risk, in many countries, new housing is still being developed on flood plains. If housing developers do not have a continuing interest in the property, the risk falls squarely on buyers who may not be aware of the risk and may find themselves in an uninsurable situation and may face heavy financial losses. The poorest sections of the population are those most at risk from climate change and, at the same time, have the fewest resources to be able to cope with the risks.

Measures to prevent and alleviate flooding have concentrated on the building of physical barriers as well as changes to agricultural use and limiting deforestation that can cause a fast runoff of rainwater into streams and rivers and exacerbate flooding downstream. Also, water holding schemes have been implemented to slow the movement of water. The design of housing may also have a role in this. For example, the covering of ground areas with tarmac or concrete will reduce ground absorption of rainwater and can increase runoff. In addition, drainage systems including the use of grey water can be engineered to reduce water runoff as well as reducing overall water consumption.

In other locations, the 'severe weather events' may involve increasing temperatures or drought, in which the problem may be the sourcing and storage of water, as described earlier. Large areas may become uninhabitable because of the difficulty of sustaining agriculture and existence there. The risk of fires may increase when rainfall decreases and the land dries out. Other problematic events may be the increased incidence of hurricanes, tornados and other excessive winds that may create structural damage. Many people already live under the threat of these and other events such as earthquakes, and housing design can have a role in alleviating structural damage and protecting residents from harm.

The changes that could potentially be caused by climate change to agriculture and the sustainability of life in certain locations could cause massive movements of populations away from areas that could become unsustainable and, therefore, put strain on existing housing and create large local shortages. Climate change may lead to massive increases in migration at local and global levels and all levels in-between that will put pressure on the resilience of housing regimes. Added to this are the difficulties caused by 'severe weather events' that may cause destruction of existing housing stock on a large scale, as has already been seen in many countries including the USA and in South-west Asia. The need in these circumstances will be for a flexible and effective emergency response and a responsive supply system that can deal with problems at a local level quickly.

Reasons for intervention

The need for government intervention to alleviate climate change, to offset its impact on individuals and communities as well as promoting a more sustainable way of living is not always accepted, particularly by neoliberal advocates of a free market. However, market incentives may not work in this instance, partly because of the incidence of costs and benefits that may fall on the society and the international community rather than the individual residents or developers. Also, the timescales may be problematic as the concern is substantially with the position of future generations that may not be involved in current decisions. As Maxton and Randers (2016, p. 74) argue, 'According to current (free market) thinking the oceans, forest ecosystems and polar ice have no economic value beyond the resources they can provide, so the cost of the damage done to them is completely ignored'. Increased resource use is at the heart of the neoliberal paradigm in housing with its social and economic pressures to invest in and consume ever-increasing amounts of housing. The increasing inequalities in consumption add to pressures to build more housing to help those who have difficulty in gaining access. Gough (2017, p. 81) argues that the inequality in housing 'spurs competitive consumption, emulation effects and excessive consumerism. It creates material aspirations that cannot be scaled to everyone in a sustainable manner. It fosters competition for 'positional goods' that is both counterproductive and unsustainable'. Klein (2014) argues that the need for a reaction to climate change 'changes everything' and is incommensurate with a market-based system. At the very least, government intervention in the market is needed to secure a response commensurate to the threat, and the housing market has a role in this. Therefore, the environmental situation is a key argument for government involvement in all aspects of housing. Without this widespread intervention in housing, it is doubtful that environmental goals could be achieved.

The wide-ranging nature of the impact of climate change means that it needs to be considered as a key element in most housing policies. So, when decisions are made about the appropriate supply system for generating a sufficient quantity and quality of housing sustainability needs to be a key constituent of discussions and a key aim of policy. Likewise, discussions about stock condition and policies to remedy defects need to bear this factor in mind. But, also, sustainability concerns and aims need to be at the heart of debates and policies about the appropriate form of neighbourhoods if environmental goals are to be achieved and global warming halted and the impacts alleviated.

Housing policy and sustainability

The embedded nature of environmental concerns means that most policy tools that are needed to achieve environmental aims are the ones that have been considered in previous chapters. The design aspects relating to the individual house may be tackled through quality controls on the standard of new buildings that, in many countries, already have a large element that relates to environmental issues. Retrofitting of the existing stock is embedded in urban regeneration policies towards existing houses and neighbourhoods. Broader sustainability and resilience issues could be embedded in urban planning processes and policies. Energy conservation measures can be achieved through subsidy for insulation measures or for more effective and sustainable heating sources.

However, each of these elements requires different forms of government intervention, and so it is likely that a wide range of methods will be necessary as there are likely to be overlaps and inconsistencies between different elements. For example, increased environmental standards in building regulations for new homes may increase the costs of housing and may have impacts on housing affordability for some groups.

The first requirement of a clear intervention designed to meet environmental objectives would seem to be a plan with clear aims and targets, given that climate change has implications across many aspects of housing and so needs to part of discussions on a wide range of topics. The importance of social practices in influencing the reaction of dwellers to environmental measures shows the importance of the discourse surrounding environmental issues. One of the major ways that government can achieve environmental goals is by influencing this discourse in the way that they define problems and the language that they use to describe issues and problems. Defining housing problems in terms of sustainability is one way they can attempt to do this.

Low-energy housing

As part of the concern about carbon dioxide emissions and their role in global warming and climate change, there has been a focus on the energy use of housing. This is unsurprising as a large part of overall energy use is associated with domestic properties (28.5 per cent in the UK according to Pérez-Lombard et al., 2008). There is also a concern from many governments about the security and resilience of energy supply that emphasises the importance of a reduction in energy use. Jones (2012) charts the move in the design of new houses from low-energy design in the 1970s that focused on reducing energy loss through higher levels of insulation and more efficient heating forms. He terms this focus as creating 'climate rejecting buildings', whereas the next phase of concern was with 'passive design' that sought to take advantage of the climate of the location through solar energy and heating. This involved using solar gain through windows and the thermal mass of the building structure to reduce the need for additional heating as well as mechanical heat recovery systems and solar energy and heating. Most recently, the concern has focused on zero-carbon or carbon-neutral housing because of the increasing concern about climate change. Zero-carbon housing makes use of the innovations pioneered in low-energy and passive house design, but seeks to reduce energy demand further and to meet the remaining use through renewable sources built into the development. The term 'carbon-neutral' has been adopted as a more useful one as it allows for the use of energy through an electricity grid at some times and input into this grid at others to achieve a neutral overall position. This allows for the evening out of variations in supply according to weather conditions or times of the day or year that are associated with some renewable energy systems such as solar or wind systems. The different emphases identified by Jones have been additive, with the one building on the other, and they have resulted in significant energy gains in new build housing (see Jones, 2012, for a review of the evidence). However, a number of issues have emerged.

The first is that the concern has been mainly with more efficiency in design and in the use of energy in the house for heating and lighting, but there has been little concern with other energy uses in the house, such as the use of appliances and communication technologies that are taking up an increasing proportion of total energy use. Second, the more sophisticated energy systems have involved intensive monitoring and control systems that have required the (sometimes intensive) involvement of the residents. This has raised the important issue of the relationship between residents and the dwelling. As Clapham (2011) has pointed out, there are many meanings and uses attached to a dwelling including those of the designer, builder and resident, and these may not coincide. Therefore, it is important for energy reduction that residents see the value of the aims and are inclined to participate in them. Otherwise the potential reductions in energy use and carbon reduction may not be achieved. This raises the important issue that the benefits of carbon reduction may accrue to the society as a whole and not all members may perceive them to be important to themselves individually. This may be especially true when there are costs involved that are borne by them and if they have difficulty in making ends meet.

Third, the incorporation of the design elements involved in carbon-neutral housing could result in increased building costs (although these may reduce in the longer term as economies of scale are achieved). The extent of any increase in prices will depend on the interaction between the supply and demand factors. Housing production systems may differ in their capacity to produce environmental standards. For example, carbon-neutral designs are often based on prefabricated or factory-built systems whereas, in many countries, there is a reliance on more labour-intensive and site-specific processes. Supply systems may vary in the incentives pursued by developers. For example, where there is a desire for short-term profit maximisation there may be less willingness to consider environmental issues than if there was also a concern with long-term usability issues. Therefore, the successful achievement of environmental aims may involve a restructuring of the housing supply system.

The impact on price may also be influenced by the willingness of consumers to pay a price premium for high environmental standards. In other words, can developers exact a price premium to offset any increased costs? This will depend on the knowledge of environmental issues by consumers and their willingness and ability to pay for improved sustainability. Any increase in environmental standards may compromise government affordability objectives by driving up prices for certain consumers.

Fourth, the design solutions used may vary in their emphasis on individual or collective solutions. For example, energy and heating systems can be based on the individual house or be organised on a community basis through district heating schemes or community energy generation through collective ownership and use of wind turbines or other renewable energy sources.

An important element in defining the extent of energy use in a dwelling is its size. The larger the floor-space the higher that energy costs are likely to be. Therefore, the recent rise in house sizes in many countries, particularly at the higher ends of the market, is a significant issue in the achievement of environmental objectives. House size is driven by the unequal distribution of housing and its positional element derived from social status. We will return to this later in the chapter, but it is clear that environmental and distributional issues are closely linked.

Retrofitting

The policy emphasis has been on new-build housing, but, as Jones (2012) points out, 80 per cent of the houses existing in the UK in 2050 have already been built. Therefore, the focus on new-built housing, whilst important as these properties will be around for a long time, will not be sufficient by itself to meet climate change targets.

Retrofitting has the possibility of making a large impact on energy use, and it may contribute to other egalitarian objectives by helping to reduce residents' costs and so helping to alleviate fuel poverty and the impact of low incomes. However, the need to 'bolt on' additional features onto an already constructed dwelling may raise many problems.

Houses will differ considerably in their flexibility, and so the possibility of making appropriate changes quickly and cheaply may vary. Policy has focused on more easily achievable features such as insulation and glazing. In some properties, insulation may be added either internally or externally through cladding and some roof spaces can be filled with insulating material. However, internal insulation may reduce the internal space, and external cladding may change the external appearance of the property, which may be a particular problem in conservation areas. External cladding of some buildings in multi-occupation such as blocks of apartments may require collective action to implement, and this may be difficult where some households lack the resources or the will to participate. Action has, also, been taken on the installation of renewable heating sources such as biomass boilers and ground source and air source heat pumps. However, there is a limit to what can be achieved through these changes, and their cost may be high as installation may be labour intensive and require changes to the fabric of the property.

The ability and willingness of occupiers or owners of property to engage in environmental improvements may depend on the impact on house value. This will partly be influenced by the ability and willingness of other consumers to pay a price premium. The existence of a gap between the cost of any improvements and the increase in price has been labelled a value gap and has been an important rationale for government involvement in the provision of grants and loans to provide the incentive for improvement.

Sustainable neighbourhoods

One of the key elements of a house is its location in relation to other places that are used by residents in pursuance of their lifestyle. The nearer these places are, the fewer resources will be used in reaching them, both in terms of time, but more importantly for our present concern, energy. There is a long tradition in urban planning, going back to Ebenezer Howard and Garden Cities, of the attempt to create urban neighbourhoods and settlements that are balanced and self-contained to avoid as far as possible commuting between different towns. Within neighbourhoods there is also a tradition of mixing land uses so that journeys to work or shops or schools are reduced to a minimum. Nevertheless, this has not always been reflected in planning policy and attempts to create sustainable settlements and neighbourhoods has not always been successful. Climate change reinforces this longstanding concern and makes the search for successful models more urgent. Case Study 11.1 reflects on the experience of the implementation of the eco-city concept in Australia.

Case Study 11.1 Eco-cities in Australia

Ecocity design refers to a suite of guiding principles and practices addressing the dominant role of urban centres in dictating unsustainable trade and development practices. It seeks to redress these unsustainable practices by embracing and celebrating the embeddedness of cities as gravitational fields of human endeavour in and as nature. On this basis, ecocity design attempts to bring the ecological footprint of the city back within the urban form, such that cities produce their own food and energy; source, treat and reuse their water; and treat and reuse their wastes. Such an ecocity would be a bioregional focal point, embedded in its hinterland, with broader trade and exchange networks predicated on socially and ecologically just practice. Complementing the ecological features of the ecocity are social, economic and political structures or processes focussing on decentralisation, participation and accessibility. Ecocity design shares many core themes with the 'alternative development' movement of earlier years, responding to reports such as that of the Club of Rome (Meadows et al., 1972; Meadows et al., 2004). It outlines and builds models for cities which can begin to counter unchecked neoliberal global capitalism by curbing consumption, reducing – if not reversing – its social and environmental impacts and fostering greater societal well-being (Daly and Cobb Jr, 1994; Register, 2001; Prugh and Assadourian, 2003). The work of Richard Register's Ecocity Builders and Paolo Soleri's Arcosanti represent key overseas education, advocacy and development efforts in this area.

In Australia, certain of the design guidelines of ecocities are being embraced by mainstream planning and development authorities. Most state planning authorities are tending towards increasing housing density, focusing new development on existing or emerging public transit and promoting mixed use development. These three traits form core tenets of new urbanism (Scheurer, 2002, 2004), though coming under fire from groups such as Save our Suburbs (SoS), which see these characteristics as threats to the suburban social and ecological fabric.

These projects highlight and reflect the conscious reinterpretation of home and neighbourhood to more directly address and nurture their political, social, economic and ecological realities. These developments attempt to take responsibility for their footprint within their form to as great an extent as possible, generating food and energy, sourcing water and reusing wastes. Where sourcing extends beyond the developments, decisions are made which reflect the groups' focus on ethical engagement, ecological sustainability and social justice. The reinterpretation of the home or neighbourhood space as food farm, waste station, water plant, 'work' space. Rather than designing 'a box to be filled with commodities' (Hayden, 1980, p. 171) which can then possibly accommodate these other activities, these projects start with an assessment of the ecological requirements and possibilities of the site and the buildings, and then build according to group desires, with an open-ended process which consciously aims to accommodate and suggest other, multiple uses and relationships.

Source: Based on Crabtree (2005).

Evaluating sustainable housing policies

There are many indicators of overall environmental impact such as the global temperature rise or the size of holes in the ozone layer. However, these general phenomena may have many causes and so evaluation of housing policy needs to be framed at a more precise level. Examples may be domestic energy use, or measures of the thermal efficiency of dwellings. Other measures may be levels of car use or distance travelled for work and social purposes.

In the rest of this book we have used subjective well-being as a criterion for evaluating housing policies, but sustainability adds an additional dimension to this. Well-being is essentially a utilitarian concept that lacks a moral and ethical dimension. Like all utilitarian concepts, it is limited in that it lacks temporal and collective dimensions. Sustainability focuses attention on the impact of policies on future generations, and this may not be picked up in well-being measures of the existing population. The temporal dimension highlights the collective dimension as dwellers are being asked to take measures that may inconvenience or have cost implications for them as individuals for the good of their children and young family members and future generations as a whole. This makes a strong case for a government prepared to take decisions and actions on the basis of the long-term interests of the society that are not recognised through market mechanisms or measures of well-being.

One of the key issues that has emerged in the discussion of the impact of environmental policies in housing relates to the social practices within the house that constitute the interactions between dwellers and the physical environment. Maller et al. (2012, p. 258) define a practice as

> a routinized behaviour involving connected elements of bodily and mental activities, objects/materials, and skills and competencies. Further, practices are interconnected with other practices (e.g. practices of cooking and practices of shopping for food) and occur within political, economic, legal, and cultural contexts of varying formality (Røpke, 2009). As Giddens (1984) proposed, each practice is shaped by the wider realm of power relations, infrastructure, technologies and society, while each practice also acts to shape these aspects of social systems.

The social practices in a house may have a profound impact on sustainability. For example, different ideas about the frequency and style of washing, which are influenced by social norms as well as by personal preferences and the marketing activities of the suppliers of washing products, can have different impacts on energy and resource use. The current design fashion in many developed countries of having a bathroom for every bedroom in a house will tend to increase use (see, for example, Maller et al., 2012). Social practices of house cleaning, as well as food preparation and cooking, will also have an influence on resource use. In their study of green renovations in Australia, Maller et al. (2012) noted that renovations to bathrooms did not challenge existing bathing social practices, but focused on achieving them more efficiently by, for example, enabling the reduction of bathing time.

> In seeking explanations we propose that households attempting to incorporate narratives of environmental sustainability into their housing renovations also contend with the accommodation of existing or future daily routines and aspirations for the ideal home. The sustainability narrative was focused on reducing consumption through improving energy efficiency, use of sustainable or second-hand goods and materials, water harvesting and recycling. Whereas aspirations for the ideal home were largely focussed on creating a home which supported household members' performance of particular social practices, such as showering, socialising and entertainment. In some instances the intersection between such narratives, aspirations and routines was compatible; in others, it was not.
>
> (Maller et al., 2012, p. 275)

As Carr and Gibson (2015, pp. 58–59) conclude, based on the Australian experience of environmental reform,

> [E]fforts towards transforming everyday practices have been met with varying degrees of success. Despite a growing enthusiasm within households for contributing to broad sustainability goals, such policies have not always solicited the intended outcomes. Smart meters do not challenge practices that householders consider non-negotiable (Strengers, 2011). Water tanks do not save as much water as predicted (Moy, 2012). Education programs emphasising that 'it's easy being green' understate the amount of domestic labour involved, and sidestep the question of who does the work (Organo et al., 2012). Residential energy consumption continues to rise, due to a combination of bigger homes containing more appliances and computer equipment, a growing population and a declining number of people per household.

As McManus et al. (2010) have observed, the average SAP rating of houses has risen steadily over the last 30 years, yet there has not been a reduction in domestic energy use, mirroring similar findings in the Netherlands by Steg and Vlek (2009) who reported that greater efficiency simply led to greater consumption. Technological fixes to homes to reduce carbon emissions may not, therefore, encourage changes in householder behaviour of the magnitude required, and whilst 'smart' homes may deliver some savings, there are considerable risks to these insofar as they detract from the need to engage wider society in the debate about lifestyle change.

The importance of social practices is shown by the major gap that exists between the technical calculations of energy use and the actual energy consumed by the same homes when occupied by real people (Sunikka-Blank and Galvin, 2012). Inhabitants seem to adjust their habits to the efficiency or standard of the building in which they are living. For example, they keep lower temperatures in inefficient houses and higher temperatures in efficient ones. Actual energy savings are thus never as great as predicted by technical calculations (Sunikka-Blank and Galvin, 2012).

Energy use in housing is influenced by the social practices that frame its consumption. An example is the reductions in size of households in many countries that increases energy use in total as many more households do not make use of the energy savings that sharing housing space can involve and use more energy per head. The nature of housing as a positional good that signals social status means that pressures to increase consumption in the form of larger houses and more gadgets and energy use are bound in closely with the current dominant discourse surrounding housing. The aim of minimisation of resource use is unlikely to be achieved if it clashes with strongly held social practices. Therefore, part of the intervention strategies of governments trying to increase sustainability may have to be aimed at changing these social practices and their embedded discourses and social norms. Social practices are sustained by 'knowledge networks' in which people exchange ideas and ways of doing things. Karvonen (2013) argues that an individual approach to sustainable renovations does not work because of the importance of contradicting social practices, but that a communal approach has a better chance of success because of the potential for transforming the social practices through knowledge exchange.

Social practices are to some extent the result of individual choices, but they are influenced by factors such as identity, belonging and social norms. Consumers of housing do not always act rationally as neo-classical economics would have us believe, but are influenced by a wide range of other factors. The nature of the housing regime also has an important influence, because it structures certain behaviours by creating systems of reward and sanction. For example, a neoliberal regime with high house price inflation rewards investment in owner-occupation and behaviour geared towards housing as a commodity. Case Study 11.2 shows the difficulties in achieving sustainability in home settings in Sweden and draws attention to the important impact of social practices based on societal norms of behaviour in reducing the impact of sustainability measures.

Case Study 11.2 Sustainability and home-related practices in Sweden

This paper explores the potential for a transition to low-impact home-related practices in high-consuming societies, with an emphasis on reducing resource intensity through minimising living space per capita, lowering living standards and promoting shared use, where residents' willingness, acceptance and support for these types of developments (and for policy measures to this end) go beyond individual changes. Such a transition will need to revolve around the co-evolution of new conventions of the "good home", exploring narratives, how they become "normalised" in practice and seeing the home as a node of multiple interlinked practices (Shove, 2003; Shove et al., 2012).

Reducing resource use in relation to spatial, material and thermal living standards is imperative to a sustainable housing development. The study presented in this paper explores perceptions among a group of residents in a "typical" multi-family owner-occupied apartment association in Gothenburg, Sweden, and their reported willingness to engage in practices and ways of living that challenge conventions surrounding home – here focused on living simpler (with a lower standard and reducing consumption), smaller, and/or sharing spaces and things. This is relevant especially in light of emerging discussions surrounding, for example, aspects of compact living and co-housing, but where the housing market is lagging behind and where it is not clear how large the interest and potential is for these types of developments.

The empirical material presented, encompassing both a questionnaire and in-depth interviews, suggests that environmental values and opinions on resource saving might be important in establishing acceptance or willingness among residents to for example lower their standards of living, but does not necessarily mean this is translated into action. Different obstacles and possibilities for adopting low-impact home practices are explored, as illustrated by interviewees' negotiations between resource saving and notions of a good, comfortable home. Changing one's dwelling situation is reliant on multiple aspects, relating to socio-structural factors and provision of alternatives as well as individual perceptions of necessity in relation to convenience. However, the rather broad openness to low-impact living that can be discerned among interviewees, and directly expressed ambitions of reducing consumption of home-related goods and prevalent discussions on the potential of living together with friends or family, offers a positive starting point for assuming that these concepts are not completely foreign to quite ordinary residents.

In line with the explored approaches to low-impact living, the interesting thing might not be whether these households are more or less environmentally oriented in general, but their approach to and critical understanding of current developments in relation to a contemporary consumer society, as expressed in the modern home and everyday life. By acknowledging the role norms (both social and regulatory) surrounding home play in challenging transitions to less resource intensive ways of living, the importance of understanding home-related practices and the negotiation of these in everyday life is underlined. Finding ways to accommodate potentially different typologies for low-impact living relies on exploring the mediation between different aspects, where living smaller or simpler might not automatically imply a willingness to live in more collaborative forms

Incentives for a more alternative development led by the market are rather low, posing an obstacle for mainstreaming of low-impact housing. Further research as well as development within the housing sector needs to address ways to create less resource intensive living environments that could appeal to a broader group of people, emphasising the role of residents in interpreting and shaping the discourse on sustainability in housing. There is a need to test and evaluate new solutions and strategies. This includes how to create environments that might allow for compromises between private and shared spaces, or compact space and resource efficient solutions, based in everyday use and the negotiation of social interactions, mundane activities, as well as reinterpreted representations of the good home. Policy should reflect and support this, emphasising explorations into alternative configurations as well as more radical reassessment of common values and norms relating to housing standards or political ideas related to the notion of a 'housing career'. Along with the market actors, media (both trade and popular media) plays a big part in proposing other narratives, where a sustainable future will mean living with less, rather than more home-related consumption – a significant challenge for developing new and existing housing within ecological limits.

Source: Based on Hagbert (2016).

The need to ask dwellers to take measures that may not help them directly and which they may not see the value of if they are unaware or doubting of environmental problems is at the heart of the difficulties faced by many programmes to make the desired impact. The chapter has shown examples where measures to achieve environmental objectives were contradictory to strongly held social practices and values that are at the heart of the neoliberal society and housing regime.

Gough (2017) has linked the welfare regimes concept to government action on climate change. He identifies climate 'leaders' as largely the social-democratic and conservative welfare regimes and the climate 'laggards' as being the neoliberal regimes. The former are likely to see economic, social and ecological values as mutually reinforcing and possess the policy tools and mechanisms to be able to pursue co-ordinated policies to achieve these aims. In contrast, neoliberal regimes 'combine weaker social policy effort with considerable levels of climate denial and less consistent attention to climate mitigation' (Gough, 2017, p. 203).

One of the major reasons that achieving sustainability objectives in housing is difficult relates to the contradictions with other objectives. For example, a commonly advocated mechanism for reducing energy use from environmentally unsustainable sources is through the pricing mechanism. Thus a 'carbon tax' could be introduced to increase the cost of energy and so discourage energy use. However, this could impact strongly on low-income households who would be increasingly in 'fuel poverty' and who spend a higher proportion of their incomes on energy even though they use proportionately less energy than higher-income groups. As Gough (2017, p. 137) concludes, 'any charge on carbon will impact more heavily on lower-income households through higher energy bills'. Compensation schemes to reward households for reducing energy consumption or installing renewable heating systems tend to favour higher-income households who can invest in the technology and are higher energy consumers. Gough (2017) identifies three pillars of carbon mitigation and argues that all three should be aggressively pursued. Raising the carbon price should be combined with direct regulation through standards (such as through building regulations) and with green investment schemes such as those to retro-fit existing housing to increase insulation and reduce energy consumption. In a review of fuel poverty, Hills (2012) concluded that measures to directly increase energy efficiency are the most effective mechanism of reducing energy use on environmental, social and distributive grounds.

Some of the policies identified in other chapters in this book to achieve redistributive and other objectives would result in increased energy use. For example, the provision of houses for homeless people and the investment in good-quality housing for low-income groups would involve increased housing investment and consumption. Therefore, if inequality and sustainability aims are to be achieved at the same time and policy is to remain in Raworth's doughnut, there needs to be redistribution between income groups and the reduction in consumption of higher-income households.

Conclusion

It is difficult to escape the conclusion that the two biggest issues facing housing policy and societies as a whole are inequality and sustainability. The urgency that faces societies in dealing with climate change and the depletion of material resources and the massive impact that they will make elevates these issues above others considered in this book. But to add to the problems, the issues are also the most intractable considered here because of their complexity and their radical challenge to existing lifestyles and social practices. Sustainability is at the heart of debates about housing policy as we have shown in this chapter that housing is implicated in many of the important issues. However, sustainability challenges both the decision-making processes of the market as well as utilitarian concepts such as subjective well-being that have been used in this book to evaluate the impact of housing policies.

The wide range of housing issues that impact on the sustainability agenda, and the need for a radical overhaul of the housing regime and the adoption of an alternative discourse and regime based on housing as a social right and the de-commodification of housing, both issues central to sustainability, mean that progress towards sustainability can be taken as a yardstick for many other important housing issues. Sustainability becomes an overall aim of housing policy that subsumes issues of over-consumption, affordability and inequality.

Perhaps more than any other area considered in this book, the existence of sustainability problems makes the case for a strong housing policy that takes into account the needs of future generations. The complexity and fundamental nature of the problems mean that they have to be confronted on a number of fronts and using the full spectrum of policies and policy tools.

Chapter 12

Neoliberalism and beyond in housing policy

The aim of this book has been to describe the current state of housing policy with examples from the six countries included here. The primary motivation in writing the book was concern about the impact of neoliberal trends in housing policy, and so the previous chapters have focused on describing and evaluating these changes. Although not all the countries considered in this book can be considered to have a neoliberal housing regime, all have exhibited moves in this direction. For example, the social-democratic regime in Sweden has moved considerably from its position since the economic crisis of the early 1990s towards the model in the Anglo-Saxon world, the UK, the USA and Australia. Argentina has many features of a neoliberal regime, although the economic situation and the level of development mean that the features of the housing regime differ from those in the developed world. The other country considered in this book, China, has what we have labelled a productivist housing regime, which is focused on the needs of the economy rather than welfare considerations. This has led to the pragmatic adoption of policies that work from other countries, such as the production of state housing for migrants into the major cities. Despite these differences, the neoliberal regime seems to be the dominant discourse in political conversations in many countries, with little competition from other ideas of what a housing regime could consist of. Therefore, in this conclusion, the aim is to provide a comprehensive assessment of neoliberal housing policy and its impacts. The previous analysis shows that these impacts are profound and have led to many undesirable elements in current housing situations. Therefore, the chapter attempts to chart a way forward to devise a housing policy that overcomes the drawbacks identified.

The chapter starts with a discussion of the ideology of neoliberalism and the concepts that have had an influence over housing policy. This is followed by a discussion of the shape that housing policy takes in implementing these concepts of neoliberalism into practice in the particular sphere of housing. We will then examine the outcomes that follow from this policy form, drawing together evidence from the previous chapters on their impact.

The aim in the first three sections is to construct an ideal type of a neoliberal model. This is based on the experience of the six countries covered in this book, but as a generalised account, it will not be reflected in full in any of the individual countries. The objective in doing this is to be able to formulate hypotheses on the nature and impact of the neoliberal discourse. The analysis is divided into three elements based on the structure used in the early chapters. Therefore, the first section examines the concepts that have shaped the implementation of the neoliberal regime that were identified in Chapter 1. The second section is an analysis of the forms of government intervention identified in Chapter 2 and applies them to the neoliberal regime to offer more detail to the emerging picture. The third section summarises the main outcomes of the neoliberal regime by collecting the main conclusions in the previous chapters on the main issues that have arisen in practice as the regime has been implemented.

The overall argument is that many of the serious housing problems encountered by the countries covered here are inherent in the neoliberal regime. Further, it is argued that these problems are not solvable within the neoliberal paradigm and that attempts to do so have been either ineffectual or counter-productive. Therefore, the chapter continues by challenging this perspective and looking for an alternative paradigm that avoids the problems and promotes a more equitable and sustainable housing regime.

Neoliberalism and housing: the concepts

In Chapter 1, neoliberalism was conceptualised as an ideological discourse. Ideologies can be seen, following Freeden (1998), to comprise shifting structures of essentially contested political concepts whose function is to simplify and control the proper meaning and structure of political thinking. In other words, to constitute the discourse that shapes understanding. The core concepts of neoliberalism are often vague and contradictory (see Birch, 2017), but following Wacquant, 2009, p. 307), neoliberalism has been taken to entail 'the articulation of four institutional logics': 'promotion, typically via economic deregulation and invariably in the name of efficiency, of markets and market-like mechanisms; welfare state retrenchment; propagation of a trope of individual responsibility and the entrepreneurial self; and an expansive and intrusive penal apparatus'. However, following Freeden (1998) attention was focused on what he terms the 'peripheral concepts' of neoliberalism that translate the core concepts into specific policies in housing. Five concepts were identified which will be used here to describe how the neoliberal discourse has been translated into specific policy using evidence from the previous chapters. The five concepts which will be covered in turn in the next sections are: privatisation, marketisation, commodification, financialisation and individualisation.

It must be borne in mind that the discourse of neoliberalism is implemented into policy through the actions of agents and institutions and that these will vary in their form and practices between countries, as we discussed in ideas of 'path dependency' in Chapter 3. Therefore, what are identified here are general trends that have been implemented to varying extents and in different forms in individual countries.

Privatisation

There are many examples of privatisation identified in the previous chapters. The most obvious and probably the most important is the 'right-to-buy' involving the selling off of state housing to tenants at a discounted price. This policy is most often associated with the UK and with Margaret Thatcher, but it has been implemented in many other countries. It is often linked with a lack of government funds for new building by the state and by no or little replacement of the houses sold, which has left room for the increase in private renting. In effect, the rental sector has been privatised. Other examples of privatisation include the refashioning of the public sector to act like private organisations. One example of this is the need for Swedish Municipal Housing Companies to act in a 'commercial' manner (Case Study 9.5). Although there is substantial disagreement about what this means in practice, it rules out the granting of precedence to vulnerable people. The case study of the Grenfell Tower fire (Case Study 5.2) in London shows how private actors had become involved in the safety regulation of buildings. Many state functions had been 'contracted out' to private organisations, including in this case management of the housing. Therefore, private agencies and private sector thinking has played an increasing role in the countries considered here with the state playing a correspondingly lesser role.

Marketisation

The involvement of private actors has often taken place through the permeation of market mechanisms into different areas of housing. Examples are the contracting out of functions such as regulation and the management of rented housing outlined above. An interesting example of marketisation is the attempts by the UK government to increase competition in the building industry by encouraging small builders in order to enable them to compete with large-scale developers that have come to dominate the industry as they are best able to cope with the fluctuations in activity that have occurred since the Global Financial Crisis (Case Study 7.2).

The privatisation of the rental system has created a market in rental housing with landlords competing with each other for tenants and so expanding the reach and scope of market activities. The market seems to be perceived as the model of a perfectly competitive neo-classical construction that will reach equilibrium and will successfully reconcile demand and supply. As we have shown in the book, and particularly in Chapter 4, this type of market does not exist in practice and indeed cannot because of the particular nature of housing.

Commodification

Increases in privatisation and marketisation have sustained and reinforced the discourse of housing as a commodity rather than as a universal right or need. Housing tenures based on meeting housing need such as the public rented sector have been reduced in scale and generally play a minor role in the housing regime (the exception here being China; see Chapter 9). The high price rises in the owner-occupied sector in many countries have reinforced a view of housing as a route to the accumulation of wealth and status rather than as primarily a place of shelter and security. The discourse of the provision of housing on the basis of ability to pay in a market is paramount.

The one exception here has been the rise of the Housing First concept (see Chapter 10) in homelessness which takes as its basis the right to appropriate, permanent and secure housing. This idea has been supported by neoliberal governments because of its proven ability to reduce state financial responsibilities compared to the 'staircase model' with its plethora of supported housing options and its discourse that gives primacy to the concept of showing the ability to sustain and deserve permanent housing. The problem with Housing First is that the neoliberal housing regime has limited the extent to which it can be implemented because of the lack of the affordable and secure housing that is essential for low-income households at risk of or subject to homelessness. Austerity in social policy has also limited the availability of the social support that is essential once secure housing has been achieved to help homeless people deal with their, often complex, problems.

Financialisation

The financialisation in housing has been well documented (see, for example, Aalbers, 2016) and has become extensive in neoliberal housing regimes as indicated by the existence of the secondary mortgage markets and the financial products that have been built up on the basis of the asset value of housing. The trend in housing is one element of what has been titled the rise of the rentier economy. As Birch (2017, p. 151) states,

> Specifically, rising asset-price inflation, resulting from increasing 'investment' in rent-earning assets (e.g. titles of existing houses) and driven by expectations of rising returns on that investment (e.g. capital gains) have meant that housing in certain parts of the USA, UK, and Canada has been transformed into an investment asset class that is simply beyond the reach of the average resident. . . . Although this example reflects the particularities of housing, it is applicable to broader trends in a political economy dominated by rentiership.

Financialisation and the rise of rentiership have been possible because of the rise of the associated trends of privatisation, commodification and marketisation that have created the space and the conditions for the asset price growth that is its basis. Financialisation has transformed housing markets and led to the many housing problems that are experienced by the countries covered here.

Individualisation

This element is connected to all the others, but is worth drawing out. Examples are the sale of council houses under the right-to-buy that have transformed a community asset, the value of the stock owned by the council that was used to invest in new collective housing at below market cost, into an individual benefit in the form of a subsidy to enable individual buyers to accumulate an individual asset. A further example is the change in the status of co-operatives in Sweden that enabled individual rather than collective asset appreciation (Case Study 4.3). Individualisation is at the heart of the other elements described above. For example, markets are based on the building blocks of a large number of individual transactions motivated by calculations of individual costs and benefits.

The neoliberal housing regime: policies and practices

A number of elements are identified here as constituting the neoliberal housing regime. It must be borne in mind again that we are constructing an ideal type from the experience in a number of countries that will not correspond directly to the situation in any single country because of the path dependent effects.

Regulation

There is a mixed record on regulation as it has increased in some areas as direct provision has reduced. The contract regime has meant that many resources are consumed in monitoring the performance of contractors in areas such as housing management or construction. Nevertheless, there have been pressures to reduce regulation in order to relieve the obligations on landlords and housing developers as examples such as the Grenfell Tower fire show (Case Study 5.2). In the UK, standards for new construction have been reduced in an effort to reduce costs to developers in the belief that this would encourage the greater level of the supply of new housing as well as help the reduction of house prices. Other forms of regulation such as the planning system are under-fire in many countries and usually blamed for any problems with the quantity of housing produced. Where rent controls exist (such as in Sweden; see Case Study 4.3), their use is constantly questioned.

Direct provision

A key part of the neoliberal housing regime is the small size of any state-provided housing sector. Where it exists, it has been constructed to mirror the private rented sector, rather than the other way around, with tenant rights in the UK, for example, being reduced so as not to compete with the private sector. Therefore, security and length of tenure have been reduced, and tenants whose incomes increase are liable to be subject to increased rents. In this regime, a public rented tenancy is not a secure tenure for life, but is perceived as a transition state for the poorest in society on their way to the private sector. State-provided rented housing has been fashioned as a temporary 'step-up' onto the housing ladder, rather than a long-term destination. It is 'social housing' in that it is restricted, wherever possible, to those who have no other place to live, because all other options in the private sector have been ruled out. But government action is focused on making these private sector options available, whether through demand-side subsidies such as housing allowance schemes in the rented sector, or support for owner-occupation.

In the neoliberal discourse, owner-occupation is seen as the most desirable tenure because of its link to financialisation and to its symbolism of a society based on asset ownership. However, as we shall see in the next section, the impact of the regime has been such that rates of owner-occupation have declined as house price appreciation has created problems of affordability for many people.

Private rental has been seen as a major tenure for those who cannot afford owner-occupation, but the tenancy conditions have been in favour of the landlords, with generally short-term tenancies and no rent control. The terms have been set to encourage investment from private sector individuals and companies, and the sector has grown in size and, in some situations, has taken on the role of social housing, by, for example, housing homeless people, underpinned by the provision of housing allowances.

Subsidies and taxation

In the neoliberal regime, the emphasis is on demand-side subsidies, either through housing allowance schemes for renters or for specific schemes to help particular marginal groups to access owner-occupation. In addition, there are usually tax reliefs for owner-occupiers, whether in the form of mortgage interest tax reliefs or exemption from capital gains tax on the sole or primary residence. Private landlords may also receive tax reliefs of various kinds. There are some direct subsidies, but these are mainly in the form of support for housing development, either in terms of financial support or in other ways such as the provision of land at below market prices or other support in the supply process.

In some countries, supply-side subsidies are retained, but these are usually small in scale and don't necessarily go to those most in need. They may be targeted at specific groups such as those in low-paid work who may have problems meeting market rent levels. If associated with right-to-buy, this may be seen as a way of enabling those households to move on to owner-occupation as their situation improves.

In most countries, the taxation of capital gains through increases in the price of land or houses is minimal. There may be some attempts to extract some return from gains due to the planning system, but these are partial and limited in scope. There is usually taxation on the use of property through a property tax, but rates are often not progressive, or at least are not proportionate as property values increase.

Information or guidance

Intervention under this heading may take many forms. There are examples in the previous chapters where governments have attempted to provide information to households conceived as consumers in housing transactions. For example, the requirements for the provision of information to prospective purchasers of housing falls into this category (see Chapter 4).

Accountability

One form of intervention is the construction of the housing regime and its institutions with its patterns of accountability and power. The two major forms of accountability in the neoliberal regime are market relations and contractualism.

The market is the dominant form in the regime and so accountability is framed in market terms. Producers and consumers interact through market transactions, and the belief is that competition will provide the stimulus for accountability. Producers will have to produce what consumers want or lose business. Consumers can choose between different products and producers or not to consume at all, thus ensuring that production meets their demands. However, we have shown in Chapter 4, and elsewhere in the book, that this theoretical construct has many flaws. Housing supply is inelastic and, in many countries, dominated by a small number of producers who are motivated by short-term profit and the optimisation of share value (see Chapter 7). Accountability is primarily to shareholders as increasing the share value has taken primacy in decision-making. Because developers make much of their profit from land transactions (therefore can sometimes increase their profit even when sales volumes fall as after the 2008 Global Financial Crisis), the consumer is a relatively minor player. The particularities of housing as a commodity and the impact this has on housing markets, means that, without corrective action, the market operates in a way that maximises producer power over the consumer.

Market transactions are contracts between a buyer and seller, and in the rental sector, the terms of consumption are defined by a lease or rental contract. Some commentators on neoliberalism have emphasised the importance of contractual relations as defining the approach and being more important than even the concern with markets (see Birch, 2017). Governments have set the terms for housing contracts and so strongly influence the balance of power between the parties. In the private rental sector, the government concern with increasing investment in the sector has meant that that the contracts between landlord and tenant have given little security to the latter and have enabled landlords to take back possession of the property easily.

In the public rented sector, the pattern of accountability has been set by the spread of contracting into the sector. In the UK, where tenants could hold their council landlord to account through democratic election processes as well as any processes of tenant participation, management and repair of the housing stock is now usually undertaken by private agencies under contract to the local authority. The terms of service are laid down in the contract terms, and accountability is primarily between the contractual parties rather than to the tenants as contracts can rarely be changed once agreed. The case study of the Grenfell Tower (Case Study 5.2) fire shows how these contractual arrangements were used by the private company and the council to avoid tenant scrutiny.

Discourse

One of the primary means by which governments intervene in housing is by setting the terms of the debate by formulating and promoting a particular discourse that sets the way of thinking about issues and problems.

The neoliberal discourse is based around the peripheral concepts considered in the last section. Owner-occupation has been consistently promoted as the tenure of choice in rhetoric as well as in financial support. The ideology of housing as a commodity rather than as a social right has generally been promoted, with acceptance and support for the capital gains to be received from the ownership of property. Therefore, the desirability of rising house prices has been promulgated despite the concomitant problems of affordability, which are often incorrectly ascribed to a shortage of new properties rather than the root causes of inequality and commodification.

This dominant discourse has also set a particular view of the status and role of state-provided rented housing that have emphasised its residual role as social housing for those who need a (temporary) helping hand to engage in the market. The view of the tenure as low status and only for vulnerable people has made proposals for expansion of the sector difficult to engage popular support.

Non-intervention

In the neoliberal regime, the concept of a 'free market' is promoted, even though we have seen how markets are constructed and maintained by the state and are rarely, if ever, competitive, because of the nature of housing as a commodity that means that the basic structures of a 'free market', such as perfect information and competitive supply structures, do not hold. Despite this, the rhetoric of market supremacy as the best way of delivering housing means that government action is characterised as 'intervention' that distorts the market and prevents it performing in the best way that leads to the optimal use of resources. Therefore, there are many examples of problems in the housing field where governments could act to achieve social objectives or to overcome particular problems, but refrain from doing so in the name of the market ideology. Examples are legion, including failure to regulate effectively the private rented sector and to alleviate affordability problems. The chapters of the book have given many examples of where government intervention in or abrogation of the market would better achieve social objectives.

Outcomes of the neoliberal regime

The preceding chapters have shown evidence of the impact of the neoliberal regime on housing issues and problems. Some of the major outcomes are discussed here as they are inherent features of the neoliberal regime.

House price appreciation

In the neoliberal regime, the dominant discourse sees housing as primarily a commodity to be traded and an individual investment for financial return. Financialisation has exacerbated this with complex financial products, based on housing, being traded in financial markets. The regime creates a situation where there is a strong interest in continuing house price inflation. This ensures that mortgage lenders can recoup the value of their loans if there are defaults; it reinforces the value of financial products. It enables developers to alleviate the risk of development, provides a profit opportunity for private landlords, leads to enlarged fee income for agencies and professionals who are involved in housing transactions, gives a financial return in terms of increasing wealth for existing owner-occupiers and entices households into the owner-occupied sector. The situation is exacerbated by the government policy of demand-side subsidies that adds to price pressures as well as the favourable tax status accorded to owner-occupiers and landowners. Ultimately, as well as the wealth generated by existing owner-occupiers, the benefits of continually rising house prices are felt by the owners of land who see the value of their asset increasing over time without them having to do anything. Housing is deeply implicated in the rise of rentiership in the world economy and is a major feature of it.

Affordability

The benefits of house price appreciation to those who are privileged enough to enjoy them are appreciable. It allows people to build up wealth to subsidise consumption or to provide security in old age. The advantages to landowners and developers are evident. But the regime leads to certain problems that impact on particular sectors of the population and particularly on the most vulnerable. The benefits to existing owner-occupiers are offset by the difficulties faced by those unable to achieve the status. Neoliberal regimes are plagued by problems of affordability for some sections of the population. The problem may be particularly acute for people living in high-pressure locations such as the major cities where prices, supported by high investment (some from outside the country) may be particularly strong. Other problems may be intergenerational with young people being priced out of housing opportunities that their parents have enjoyed.

Volatility

The extent of house price volatility varies between different countries, but it seems to be an integral part of the neoliberal housing regime. The liberalisation and globalisation of finance, as well as the financialisation of housing itself, have created a situation where variations in economic performance can be reflected in changes in house prices, levels of new building, and of housing transactions. The stabilisation mechanisms that governments have used in the past to iron out these fluctuations have been rendered ineffective or impractical by financialisation as was shown in Chapter 8.

Housing shortages

The neoliberal regime is characterised by a housing supply that is dominated by a few large companies that are motivated by short-term share price considerations and make major gains from land transactions. They can indulge in anti-competitive practices protected by high barriers to entry and difficult operating conditions for small companies. Production can be maintained to cope with risk and to increase profits. The result is inelasticity of supply that does not respond effectively to demand pressures. When this is coupled with the pressures to increase consumption through status considerations fuelled by inequality, the result is shortages of accommodation in particular places and for particular people, especially those on low incomes.

Inequality

Neoliberal countries have relatively high levels of income and wealth inequality. The housing situation is both a reflection of this overall situation as well as a contributor to it. The differential benefits of house price increases have helped some segments of the population to receive benefits denied to other, usually more vulnerable and poorer sections, thus exacerbating the inequality. At the heart of this is the increasing importance of rentiership that allows those with capital to gain income at a greater rate than is available to those relying on income from their labour (Piketty, 2014).

Inequality is reflected in housing outcomes. For example, the distribution of housing space in the UK has become more unequal (Dorling, 2014). At the same time, the situation of those excluded from owner-occupation such as young people and renters have seen their housing situation worsen relatively. Increased inequality is also reflected in increased segregation of low-income and other disadvantaged groups as was shown in Chapter 6.

Inequality in housing is particularly important for a number of reasons. One is the access that housing provides to other life chances such as health and employment opportunities. But also important is the positional element of housing. The situation of others we compare ourselves with has an important impact on our well-being and health. The wider the inequalities the more impact this has. There is substantial evidence that inequality damages the health (as well as the economic situation) of the society taken as a whole, and there is emerging evidence that this is also true of housing (see Chapter 8).

Homelessness

There is some evidence that neoliberal regimes experience higher levels of homelessness (see Chapter 10). At the same time, policies that are most effective in dealing with homelessness, such as Housing First, are difficult or impossible to implement on a wide scale because of the lack of affordable housing for participants to access. Because of its profound impact on the lives of those experiencing it, homelessness is the most extreme illustration of housing inequality.

Sustainability

It was argued in Chapter 11 that markets in general, including the housing market, do not reflect well issues such as sustainability that involve the interests of future generations. Markets, even when working as the neo-classical theory postulates, only reflect the current interests of consumers and producers. Governments of all persuasions have seen sustainability as an issue for government action, even though the nature, scale and extent of this action has varied considerably. In neoliberal housing regimes, where commodification and marketisation are key elements, there are strong forces towards increasing housing consumption, especially given the problems in providing adequate housing for vulnerable groups in an unequal society.

The analysis above explains why progress in implementing measures to respond to climate change and the shortage of some materials has been so limited and halting. For example, regulation to achieve these aims in the UK has been cut back when it was seen as leading to increased costs for housing producers. A main barrier to sustainability is the consumerism driven by commodification that leads to demands for increased consumption. The existence of inequality also leads to demands for increased consumption because of positional factors as well as the need to provide levels of consumption for the poorest of the population.

Societal well-being

There is some evidence, although scant at this time, that neoliberal regimes result in less overall well-being than other regimes where inequality in general and in housing in particular is less. This mirrors the findings on many social problems as reflected in the work of social epidemiologists (see, for example, Wilkinson and Pickett, 2010). This finding probably stems from the positional nature of housing, reflecting the importance of identity and status in housing satisfaction. This is a much under-researched field in housing studies and deserves more attention from housing researchers.

In addition, the problems of inequality and homelessness show that the neoliberal regime has resulted in the restriction of the capabilities for many vulnerable people. Homelessness results in disadvantage in many other areas of life, and the high level in neoliberal regimes means that these problems are of an unacceptable scale. Inequality results in many vulnerable people living in poor and insecure housing and neighbourhoods that also impacts on their capabilities.

General

The outcomes identified above are endemic to the neoliberal regime and show that the regime is unable to meet the success criteria that are often claimed for it. The affordability, supply and volatility problems mean that the size of the owner-occupied sector is declining in many countries with little sign of this decline being reversed. The model is based on the importance of house price appreciation, but this limits the spread of owner-occupation and the short-term disposable incomes of those who enter the high-cost sector. Neoliberal housing markets do not function well, as the problems with supply inelasticity and housing shortages show. The nature of housing and its importance as a positional good mean that the benefits cannot be widely shared between generations as the increasing volume of provision means that some of the advantages enjoyed by early pioneers are eroded by the mass availability (see Chapter 8). Meanwhile many people are constrained to living in unregulated rental sectors with insecurity and high rental payments. The impact of the neoliberal housing regime is amplified by the application of neoliberalism in other fields such as employment and welfare benefits as well as in policies towards inequality in general. Combined, these policies have resulted in high levels of inequality and poverty that have created problems of social cohesion and economic growth. Changes to the labour market have created insecurity for many people that makes it difficult for them to access and sustain owner-occupation.

The contention here is that these impacts are endemic to neoliberalism and so constitute a challenge to the discourse that promises impacts that are not borne out in practice such as efficient and responsive housing markets, the spread of the benefits of owner-occupation, and competitive and attractive rental markets.

Varieties of residential capitalism

Before focusing on the outcomes of the neoliberal housing regime, it is worthwhile briefly discussing other elements of a neoliberal society. It was highlighted in Chapter 1 that many other elements of society, as well as public policies in other fields, impact on housing outcomes. It is not within the scope of this book to cover these in any detail, but it is worthwhile mentioning some here. Financialisation and the rise of 'rentier capitalism' have been mentioned in earlier sections as they relate directly to the housing regime. Three other related features of neoliberal societies stand out. One is the functioning of the labour market. Neoliberalism has been associated with a 'flexible' labour market that has seen increases in self-employment, what has been called 'the gig economy' where people take on specific and short-term contracts, casual and zero-hours contracts, fixed-term labour contracts and increasing job insecurity. All of these impact on the level and security of earnings of households. One of the reasons for the decline in owner-occupation in some neoliberal regimes has been the contradiction between the need to repay regular, large sums to financial organisations and the insecurity and uneven levels of household income.

The second factor is the growing inequality of incomes and wealth. This is discussed in Chapter 8 in relation to housing costs and subsidies, but inequality impacts on many areas of life, many of which also impact on housing outcomes.

The third factor is the general pressures on public services caused by reduced or low rates of financing. In areas such as homelessness, the shortage of effective social services and health care has added to the extent of the problem and made solutions more difficult to achieve. Changes to general welfare benefits can have profound impacts on the ability of households to meet housing costs whilst also managing to pay for other essentials of life such as food and heating.

The key point is that neoliberalism in housing has been accompanied by the imposition of neoliberal values and policies in other spheres that have impacted on housing. This raises the important question of how much progress can be made in dealing with the housing outcomes identified without tackling the broader issues.

Prospects for change within neoliberalism

It must be stressed that individual countries have the problems identified with the neoliberal regime to varying degrees and they take a different form, depending on the specific nature of the housing regime. Nevertheless, the problems are endemic in the neoliberal regimes, and so it is worthwhile examining how they could be overcome. Is it possible to change these impacts through policy change within this regime or is more radical change required?

Within the individual neoliberal housing regimes themselves various attempts have been made to deal with the problems, many of which have made the problems worse. To take the UK as an example, one response has been to increase demand-side subsidies for owner-occupation which have helped some individual households, but have resulted in increased prices in general and so have made the problem worse. For rented housing, the response has been to reform welfare benefits to reduce public expenditure, but this has created real hardship amongst those affected and not had any appreciable impact on general rent levels. Many private landlords have responded by not letting to benefit claimants and moving their business to higher-income tenants where the returns are greater. There has been much focus on housing supply in the belief that there is a shortage of housing which is responsible for the price increases. This belief disintegrates when it is recognised that there is more housing space per person than ever before in the UK. In addition, new build in any year is a very small proportion of total housing stock, and so it would take sustained production, over a long time-period, to make any appreciable impact on house prices. The key issue is not any overall shortage but the unequal distribution of the existing stock. Any new build will not help this if it is not aimed at those who are most in need and is merely taken up in the consumption of more space by those with the most resources.

Some clues on how to deal with the problems can be taken from experience in the countries that have progressed less far down the neoliberal road such as Sweden and China. This is not to say that they do not experience these problems as both have issues of affordability in some locations and have experienced house price inflation. In China, the response in the major cities where the affordability problems are most acute has been to build more public rented housing directly aimed at those in most acute need. Sweden has a legacy of public housing that provides refuge even though its use has been circumscribed by the neoliberal reforms to the mission of the municipal housing companies.

So, some problems in the UK could be overcome by the building of state-provided rented housing, the abolition of demand subsidies for owner-occupation and tax changes to owner-occupation to help reduce house price inflation. However, these changes would require a major turnaround in government direction and challenge key tenets of the neoliberal approach, and it seems unlikely that they would receive public support given the strength of the dominant discourse. Also, it is likely that more radical changes would be needed to make any impact on the problems. Some neoliberal regimes have elements of these policies but still suffer from many of the identified problems. Therefore, it is worth standing back and examining what an alternative regime would look like to set the overall direction of change.

Towards an alternative regime?

It may be easier to say what an alternative regime would not be than to specify its form. Certainly, it would have to reverse the trends towards privatisation, commodification, marketisation, financialisation and individualisation that have underpinned the neoliberal regime. The fundamental concept is the view that housing is a universal and basic human right rather than primarily a traded commodity. The aim is for housing to be of value for what it affords households rather than its value as a store of wealth or as a price in a market transaction. The other two main fundamental values are ones of equality and sustainability with the well-being of dwellers at its core.

An alternative housing regime should, at the very least, be structured in an attempt to offset any inequality in the wider society. It is likely that the less the inequality in the wider society the more successful will be the housing regime. The reduction of inequality is important in order to improve the housing possibilities for the more vulnerable households such as those in danger of, or experiencing, homelessness. But it is also important in countering the positional aspects of housing. The demand for ever-greater consumption is fuelled by concerns about status compared with others. As we outlined earlier, well-being in housing is greater in societies where there is less inequality, probably for this reason.

The importance of environmental sustainability was articulated in Chapter 11. Housing regimes should be structured in order to achieve general environmental goals in the usage of scarce materials and in energy consumption.

Therefore, the four principles underlying an alternative housing regime are housing as a right, well-being, equality and sustainability. So, what would a regime built on those foundations look like, and what policy instruments would be appropriate?

An alternative housing regime: policies and practices

Overcoming the problems of the neoliberal regime will demand action in a number of spheres. For example, inequality and sustainability are wide-ranging issues that need tackling from a wide variety of angles. In these, as in other areas, action in housing could be undermined or even rendered completely irrelevant by more general issues. For example, trying to decrease inequality in the housing regime is not likely to succeed if overall inequality increases. Likewise, action to overcome financialisation in housing will be more successful if it is part of a general attack on rentiership. Bearing this in mind, the chapter now focuses on what action can be taken in housing and what an alternative housing regime based on rights, equality, sustainability and well-being would look like.

Regulation

There has been a reduction in regulation in many areas in the neoliberal regime, but three areas stand out for a renewed impetus. The first is regulation of the private rented sector. In many neoliberal regimes, regulation has favoured landlords and developers, in the belief that this would encourage greater supply. However, the result has been increased tenant insecurity and a very large demand-side subsidy cost as rents have increased, particularly in areas of high housing stress. State-provided rented housing has been run down in many regimes and the terms of tenancy set so as not to compete with the private sector in tenant conditions. In an alternative regime that puts the well-being of households first, tenancy conditions need to be improved, and the private sector should be led by the public and not the other way around. Long-term secure and affordable tenancies seem to be the minimum that a well-being-based regime would include, together with effective rights to repairs. The goal of a common tenancy between the public and private sectors is an important one.

The second area of increased regulation is in sustainability as argued in Chapter 11. There is a need for more extensive and effective regulation around new-build housing as well as programmes to retro-fit existing housing. The land use planning system is an important element in achieving the balance between sustainability and other objectives and has been the subject of restrictions in the neoliberal regimes that should be reversed.

The third area is the regulation of mortgage lending and the secondary mortgage markets. The financialisation of housing has brought with it insecurity, volatility and house price inflation. The housing mortgage market and the financial products that exist on the back of it need to be reconstructed to match societal objectives in housing of affordability and stability. Governments construct markets, and there is an urgent need to reconstruct this one.

Direct provision

The direct provision by the state or state agencies of housing for rent achieves a number of objectives. These were described in Chapter 9 but are worth repeating here: it helps to increase the quantity and quality of housing produced; it provides a way of evening out cycles in housebuilding activity; if built with supply-side subsidies, it provides an affordable and secure home for low-income households; it offers an alternative to private renting and owner-occupation, thus helping to relieve pressure on those sectors; it provides downward pressure on rent levels as the state can take advantage of low-cost borrowing and economies of scale in construction; it can provide a mechanism of collective capital gain that can be used to fund new house-building at below market rates and to cross subsidise between expensive and cheaper buildings constructed at different times; it helps to reduce homelessness and provides a resource that can be used for schemes aimed at dealing with homelessness when it occurs such as Housing First. It can embed principles of sustainable construction and use. Given these advantages, why would any government not have a programme of newly built state-provided rented housing?

The reason for its unpopularity in neoliberal discourse and the neoliberal regimes, is that it directly contradicts the commodification of housing as well as the other peripheral concepts of the neoliberal regime. In contrast, it reinforces the discourse of housing as a basic need and right. Therefore, policies should include a suspension of privatisation policies such as the Right-to-buy, which should only be continued if there is a strict policy of a like for like replacement of any public rented housing lost to the tenure. Discounts should be abolished or kept at a minimum as there is little justification for turning a community asset into individual, personal gain. The same principles should apply to regeneration schemes that reduce the amount of affordable housing available and forcibly move people away from their neighbourhoods. The cornerstone of the direct provision agenda is a programme for new building of state-provided rented housing that is accountable to residents and local communities. The particular mechanism for achieving this may vary from country to country depending on the current institutional structure.

Chapter 9 showed the pitfalls that can occur in attempts to create a successful state-provided rented sector. There needs to be high-standard and well-designed accommodation scattered in good neighbourhoods and let at affordable rents. The neoliberal conception of a residual sector only for those unable to own or rent privately is unlikely to enhance the well-being of its residents and provide the secure home that they require. Considerable thought needs to go into the design of the sector to ensure efficiency of repairs and management and accountability to residents to ensure that services are appropriate and to avoid major crises such as the Grenfell Tower fire.

Subsidies and taxation

There are two major aims in this category. The first is to prevent, or at least to control to an acceptable level, house price appreciation in order to achieve the main aim of increasing affordability. The first requirement here is to remove demand-side subsidies for owner-occupation that help recipients in the short-term, but make affordability worse in the long-term by increasing prices. Taxes on housing transactions should also be abolished as they hinder effective market operations. Taxation should be applied to owner-occupation to bring it into line with the tax treatment of other assets. Therefore, there should be some form of taxation of capital gains applied to the primary residence. Taxes on consumption in order to pay for infrastructure and other public services should be progressive in order to reduce incentives for greater housing consumption and provide a mechanism to move towards a more efficient and equitable distribution of housing space. For example, it may provide a financial incentive for some older people to move to smaller accommodation and so free up space for young families.

In addition, there is a strong case for tackling price increases at a fundamental level by accepting the unique position of land as a resource and applying a land value tax to ensure its most effective use and to reduce the rentier returns from land ownership.

Finally, the use of demand-side subsidies in rental housing needs to be reduced because of the negative incentive effects it provides for low-income households as well as the impact on rent levels and land values. It was argued above that the state-provided rented sector should be regarded as the market leader, and here there should be more use of supply-side subsidies to place downward pressure on rent levels. Private sector rents can then be controlled in relation to this. This approach could save substantial government expenditure on housing allowance programmes as well as overcoming their disincentive effects and putting downward pressure on rent levels and land prices. The control of private sector rents should impact on land prices in the medium term and may not impact too much on profit levels of developers and landlords.

Information and guidance

It was argued in Chapter 4 that there are information asymmetries in the housing market and that government action could help to overcome these through the mechanisms designed to provide appropriate information to consumers. An example is the information provided to prospective purchasers of owner-occupied houses or the information on the rights and obligations of tenants. The alternative regime could see these information requirements established as a set of housing rights and a mechanism devised for their effective implementation and enforcement.

Accountability

The implementation of a regime based on housing as a right is dependent on the existence of a framework of practical and enforceable rights. A major task of government is to create the mechanisms by which this can be implemented. Therefore, public and private landlords should be accountable to their tenants for the implementation of their rights, just as tenants should be held to account for their responsibilities. The providers and managers of public rented housing should be accountable to their tenants and other local residents and lending agencies more accountable to central government for their lending policies.

Discourse

The strength of the neoliberal discourse has meant that many of the policies outlined above may be seen as outrageous in that they contradict the taken-for-granted assumptions that have been promoted as 'reality'. Therefore, a major task is to rewrite that definition of reality and to draw attention to another way of seeing things. The battle of ideas will be crucial in moving towards a housing regime that serves the population more effectively and equitably.

The importance of the discourse surrounding housing was shown in Chapter 11 in the debate on the social practices that can enable or frustrate the achievement of sustainability goals. The desire to enact sustainable practices often conflicted with social and personal norms about the 'good' and preferred lifestyle. Achievement of sustainability objectives needs the creation and acceptance among the population of necessary social practices in housing and in other spheres.

Non-intervention

The predilection toward non-intervention has to be superseded by a government willingness to intervene when the outcomes and processes of the housing regime are not those desired. This involves government taking responsibility for these housing outcomes and not hiding behind a belief in market supremacy that the preceding chapters show does not exist in the deeply dysfunctional neoliberal housing markets.

Making it happen

The alternative regime outlined above offers the hope of avoiding the problems inherent in the neoliberal regime, but the key question concerns the possibility of implementing this vision. If the vision is in place, change can be incremental if it leads in a consistent direction, but key elements, such as the moderation of house price increases will take radical change before the advantages appear. A revolutionary approach may offer benefits more quickly.

As Birch (2017, p. 185) argues:

> [T]he only way I can see us getting out of the current neoliberal system crisis is through a massive devaluation of housing and other assets (e.g. fossil fuel reserves). Current fears about the future in light of events in 2016 pale in comparison to the potential anger that could result from even suggesting that we collectively devalue every-one's housing assets. If any government implemented such a plan, it would likely end in a popular uprising of the sort to make a zombie apocalypse seem tame in comparison. However, leaving societies and economies as they are – highly unequal, rising debt burdens, declining prospects for younger generations, etc. – is no solution either.

In Chapter 2 the concept of a discursive approach to policy-making was introduced which focused on three stages. The first is 'language games' that involved the battle of ideas and discourses in defining the issues and problems. The second is the 'coalition-building strategies' that actors use to further their position. The third stage is the analysis of the power relationships in which actors seek to negotiate and impose their views in a conflictual arena.

The need for an alternative to the neoliberal discourse has already been emphasised. A discourse that can be used to challenge the dominance of neoliberalism has been outlined above. The existence of the problems identified in the outcomes of the neoliberal paradigm is important and the knowledge that they are endemic in the approach is a strong reason to discard this framework of thought. It is readily apparent that governments in many countries have not been able to overcome these problems because of the self-adopted straitjacket of neoliberalism. The establishment of successful examples of new approaches, such as sustainable settlements, or Community Land Trusts, is important in giving practical shape to abstract ideas and principles and shows that, with the right support, they can work.

The second stage draws attention to the political support that any challenge to neoliberal housing regimes will have to face. There are strong vested interests in the status quo such as landowners and developers. The benefits have been shared with the many existing owner-occupiers who have made large capital gains from their ownership of an asset that has increased substantially in value. But existing owner-occupiers have sons and daughters and other family members who have become excluded from the sector because of affordability problems. The recent reductions in the size of the owner-occupied sector in many countries changes the electoral arithmetic for political parties to consider. All people can see the plight of homeless people on the streets and the inequality this illustrates. The problems of private renters are very evident in many countries. The problems of the neoliberal regime are severe and profound enough to provide the ammunition for an alternative to be considered and adopted, and have left many groups and individuals to face hardship and disadvantage. Social movements of those affected have been based on these problems and issues, such as resistance to urban regeneration schemes that threaten neighbourhoods and reduce much needed state-provided rented housing, or squatter groups formed by those unable to access affordable housing. These, sometimes disparate, groups need to be co-ordinated and linked into political parties and processes.

The neoliberal regime has the support of many powerful interests. It has been shaped in the interests of landowners, developers, landlords, property professionals and financial institutions as well as existing house owners and so they are likely to resist change. The neoliberal housing regime is also tied in with similar structures in other spheres and the wider economy. Housing is just one of the many features of 'rentier capitalism'. Therefore, change in housing is probably linked and perhaps dependent on similar change in other spheres. But housing is an area where the problems are very apparent and impact strongly on people's lives. Therefore, it could be a strong lever for more general change. In Chapter 1 the concept of a critical juncture, where radical change is possible, was introduced. It remains to be seen whether we are at such a time now.

Bibliography

Aalbers, M. B. 2008. The financialization of home and the mortgage market crisis. *Competition and Change*, 12(2), 148–166.
Aalbers, M. B. 2015. The great moderation, the great excess and the global housing crisis. *International Journal of Housing Policy*, 15, 43–60.
Aalbers, M. B. 2016. *The financialisation of housing: A political economy approach.* Abingdon, Routledge.
Adamson, D. & Bromiley, R. 2008. *Community empowerment in practice: Lessons from communities first.* York, Joseph Rowntree Foundation
Allen, C. 2008. *Housing market renewal and social class.* London, Routledge.
Allen, J. 2004. *Housing and welfare in Southern Europe.* Oxford, Blackwell Publishing.
Ambrose, B. W. & Peek, J. 2008. Credit availability and the structure of the homebuilding industry. *Real Estate Economics*, 36, 659–692.
Andrews, D. 2010. Real house prices in OECD countries: The role of demand shocks and structural and policy factors. OECD Economics Department Working Papers No 831. Paris: OECD.
Apgar, W. & Baker, K. 2006. *The evolving homebuilding industry and the implications for consumers.* Cambridge, MA, Harvard University Press.
Arthurson, K. 2010a. Operationalising social mix: Spatial scale, lifestyle and stigma as mediating points in resident interaction. *Urban Policy and Research*, 28, 49–63.
Arthurson, K. 2010b. Questioning the rhetoric of social mix as a tool for planning social inclusion. *Urban Policy and Research*, 28, 225–231.
Atkinson, A. 2015. *Inequality: What can be done?* Cambridge, MA, Harvard University Press.
Atkinson, A. & Bridge, G. (eds.) 2005. *Gentrification in a global context: The new colonialism.* London, Routledge.
Atkinson, R. & Kintrea, K. 2002. Area effects: What do they mean for British housing and regeneration policy? *International Journal of Housing Policy*, 2, 147–166.
Auyero, Javier. (2012). *Patients of the state: The politics of waiting in Argentina.* Durham: Duke University Press.
Ball, M. 2012. Housebuilding and housing supply. In: Clapham, D., Clark, W. & Gibb, K. (eds.) *Sage handbook of housing studies.* London, Sage.
Ball, M. 2013. Spatial regulation and international differences in the housebuilding industries. *Journal of Property Research*, 30, 189–204.
Barker, K. 2004. Delivering stability: Securing our future housing needs. Final Report: Recommendations. London.
Barker, K. 2014. *Housing: Where's the plan?* London, London Publishing Partnership.
Bengtsson, B. 1994. Swedish rental policy: A complex superstructure with cracking foundations. *Scandinavian Housing and Planning Research*, 11, 182–189.

Bengtsson, B. 2001. Housing as a social right: Implications for welfare state theory. *Scandinavian Political Studies*, 24(4), 255–275.
Bengtsson, B. 2004. Swedish housing corporatism – a case of path dependence? Paper presented at the ENHR 2004 conference, 'Housing: Growth and regeneration'. Cambridge, UK.
Bengtsson, B. 2008. *Why so different?: Housing regimes and path dependence in five Nordic countries*. Gävle, IBF.
Bengtsson, B. & Ruonavaara, H. 2010. Introduction to the special issue: Path dependence in housing. *Housing, Theory and Society*, 27, 193–203.
Birch, K. 2017. *A research agenda for neoliberalism*. Cheltenham, Edward Elgar.
Blair, R. & Sokol, D. (eds.) 2015. *The Oxford handbook of international anti-trust economics*. Oxford, Oxford University Press.
Blunden, H. & Drake, G. 2016. Homelessness, the "housing first" approach and the creation of home. *In:* Dufty-Jones, R. & Rogers, D. (eds.) *Housing in 21st century Australia: People, practices and policies*. London, Routledge.
Blyth, M. 2013. *Austerity: The history of a dangerous idea*. Oxford, Oxford University Press.
Boddy, M. 1980. *The building societies*. London, Macmillan.
Bourdieu, P. 1970. The Berber house or the world reversed. *Information (International Social Science Council)*, 9, 151–170.
Bourdieu, P. 1984. *Distinction: A social critique of the judgement of taste*. London, Routledge.
Bowie, D. 2017. *Radical solutions to the housing supply crisis*. Bristol, Policy Press.
Bramley, G. & Fitzpatrick, S. 2018. Homelessness in the UK: Who is most at risk? *Housing Studies*, 33, 96–116.
Brown, R. 2017. *The inequality crisis: The facts and what we can do about it*. Bristol, Policy Press.
Byrne, D. S. 1986. *Housing and health: The relationship between housing conditions and the health of council tenants*. Aldershot, Gower.
Caincross, L., Clapham, D. & Goodlad, R. 1997. *Housing management, consumers and citizens*. London, Routledge.
Caldera, A. & Johansson, A. 2013. The price responsiveness of housing supply in OECD countries. *Journal of Housing Economics*, 22, 231–249.
Calvani, T. & Siegfried, J. 1979. *Economic analysis and anti-trust law*. Boston, Little, Brown and Co.
Carmona, M., Carmona, S. & Gallent, N. 2003. *Delivering new homes*. London, Routledge.
Carr, C. & Gibson, C. 2015. Housing and sustainability: Everyday practices and material entanglements. *In:* Rogers, D.-J. & Dufty-Jones, R. (eds.) *Housing in 21st century Australia: People, practices and policies*. Abingdon, Routledge.
Carter, M. 1995. *Out of sight: London's continuing B and B crisis*. Crisis, London.
Castles, F. G. 1998. *Comparative public policy: Patterns of post-war transformation*. Northampton, MA, Edward Elgar Publishing.
Cheshire, P. & Sheppard, S. 1998. Estimating the demand for housing, land, and neighbourhood characteristics. *Oxford Bulletin of Economics and Statistics*, 60, 357–382.
Chiu, R. L. H. 2004. Socio-cultural sustainability of housing: A conceptual exploration. *Housing, Theory and Society*, 21, 65–76.
Christophers, B. 2013. A monstrous hybrid: The political economy of housing in early twenty-first century Sweden. *New Political Economy*, 18, 885–911.
Clapham, D. 1995. Privatisation and the East European housing model. *Urban Studies*, 32, 679–694.
Clapham, D. 2003. Pathways approaches to homelessness research. *Journal of Community & Applied Social Psychology*, 13, 119–127.
Clapham, D. 2005. *The meaning of housing: A pathways approach*. Bristol, Policy Press.
Clapham, D. 2006. Housing policy and the discourse of globalization. *European Journal of Housing Policy*, 6(1), 55–76.

Clapham, D. 2010. Happiness, well-being and housing policy. *Policy & Politics*, 38, 253–267.
Clapham, D. 2011. The embodied use of the material home: An affordance approach. *Housing, Theory and Society*, 28, 360–376.
Clapham, D. 2014. *Regeneration and poverty in Wales: Evidence and policy review*. York, Joseph Rowntree Foundation.
Clapham, D. 2017. *Accommodating difference: Evaluating supported housing for vulnerable people*. Bristol, Policy Press.
Clapham, D. 2018. Housing theory, housing research and housing policy. *Housing, Theory and Society*. forthcoming.
Clapham, D., Foye, C. & Christian, J. 2017. The concept of subjective well-being in housing research. *Housing, Theory and Society*, 1–20.
Clapham D., Hegedus, J., Kintrea, K., & Tosics, I. with Kay, H. (eds) (1996) *Housing privatisation in Eastern Europe*. Westport, CT, Greenwood Press.
Clapham, D. & Kintrea, K. 1987. Importing housing policy: Housing co-operatives in Britain and Scandinavia. *Housing Studies*, 2, 157–169.
Clapham, D., Kintrea, K. & Macadam, G. 1993. Individual self-provision and the Scottish housing system. *Urban Studies*, 30, 1355–1369.
Clark, J. & Kearns, A. 2012. Housing improvements, perceived housing quality and psychosocial benefits from the home. *Housing Studies*, 27, 915–939.
Clark, W. 2009. Changing residential preferences across income, education and age: Findings from the multi-city study of urban inequality. *Urban Affairs Review*, 44, 334–355.
Cloke, P., Milbourne, P. & Widdowfield, R. 2000. Change but no change: Dealing with homelessness under the 1996 Housing Act. *Housing Studies*, 15, 739–756.
Coiacetto, E. (2006). Real estate development industry structure: Consequences for urban planning and development. *Planning Practice and Research*, 21, 423–441.
Coiacetto, E. 2009. Industry structure in real estate development: Is city building competitive? *Urban Policy and Research*, 27, 117–135.
Cole, I. & Furbey, R. 1994. *The eclipse of council housing*. London, Routledge.
Colic-Piesker, V., Ong, R. & McMurray, C. 2010. *Falling behind: The growing gap between rent and rent assistance 1995–2009*. Melbourne, AHURI.
Communities and Local Government. 2010. *Local decisions: A fairer future for social housing*. London, Communities and Local Government.
Crisis and Shelter. 2014. *A Roof over my head: the final report of the Sustain project*. London, Crisis and Shelter.
Czasny, K., Feigelfeld, H., Hajek, J., Moser, P. & Stocker, E. 2008. *Housing satisfaction and housing conditions in Austria in a European comparison*. SRZ, Vienna.
Dalton, T. 2012. First home-owner grants. *In:* Smith, S. (ed.) *International encyclopaedia of housing and home*. Oxford, Elsevier.
Dalton, T., Pawson, H. & Kath, H. 2015. *Rooming house futures: Governing for growth fairness and transparency*. Melbourne, AHURI.
Davies, W. 2017. *The limits of neoliberalism*. London, Sage.
Dean, J. & Hastings, A. 2000. *Challenging images: Housing estates, stigma and regeneration*. Bristol, Policy Press.
Deaton, A. 2013. *The great escape: Health, wealth and the origins of inequality*. Princeton, NJ, Princeton University Press.
Department of the Environment, 1994. *Access to local authority and housing association tenancies: A consultation paper*. London, DOE.
Després, C. 1991. The meaning of home: Literature review and directions for future research and theoretical development. *Journal of Architectural Research*, 8, 96–155.
Doling, J. F. 1997. *Comparative housing policy: Government and housing in advanced industrialized countries*. London, Macmillan.

Dolowitz, D., with Holme, R., Nellis, M., & O'Neill, F. 2000. *Policy transfer and British social policy: Learning from the USA?* Buckingham, Open University Press.

Dorling, D. 2014. *All that is solid: How the great housing disaster defines our times and what we can do about it.* London, Penguin Books.

Dufty-Jones, R. & Rogers, D. 2016. *Housing in 21st-century Australia: People, practices and policies.* London, Taylor and Francis. Kindle Edition.

Dunleavy, P. 1981. *The politics of mass housing in Britain 1945–75.* Oxford, Clarendon Press.

Durnova, A., Fischer, F. & Zittoun, P. 2016. Discursive approaches to public policy: Politics, argumentation, and deliberation. *In:* Peters, B. G. & Zittoun, P. (eds.) *Contemporary approaches to public policy: Theories, controversies and perspectives.* London, UK: Palgrave Macmillan.

Elsinga, M. & Hoekstra, J. 2005. Homeownership and housing satisfaction. *Journal of Housing and the Built Environment*, 20, 401–424.

Esping-Andersen, G. S. 1990. *The three worlds of welfare capitalism.* Cambridge, Polity Press.

Etzioni, A. 1986. Mixed scanning revisited. *Public Administration Review*, 46, 8–14.

Fischer, F. & Forester, J. 1993. *The argumentative turn in policy analysis and planning.* Durham, NC, Duke University Press.

Fisher, P. & Collins, A. 1999. The commercial property development process. *Property Management*, 17, 219–230.

Fitzpatrick, S. 2005. Explaining homelessness: A critical realist perspective. *Housing, Theory and Society*, 22, 1–17.

Fitzpatrick, S. 2012. Homelessness. *In:* Clapham, D., Clark, W. & Gibb, K. (eds.) *Sage handbook of housing studies.* London, Sage.

Fitzpatrick, S. & Pawson, H. 2007. Welfare safety net or tenure of choice? The dilemma facing social housing policy in England. *Housing Studies*, 22, 163–182.

Fitzpatrick, S., Pawson, H., Bramley, G., Wilcox, S. & Watts, B. 2016. *The homelessness monitor 2016.* London and York, Crisis and Joseph Rowntree Foundation.

Fitzpatrick, S. & Pleace, N. 2012. The statutory homelessness system in England: A fair and effective rights-based model? *Housing Studies*, 27, 232–251.

Fitzpatrick, S., Quilgars, D. & Pleace, N. (eds.) 2009. *Homelessness in the UK: Problems and solutions.* Coventry, Chartered Institute of Housing.

Fitzpatrick, S. & Stephens, M. 1999. Homelessness, need and desert in the allocation of council housing. *Housing Studies*, 14, 413–431.

Forrest, R. & Kearns, A. 2001. Social cohesion, social capital and the neighbourhood. *Urban Studies*, 38, 2125–2143.

Foucault, M. 2008. *The birth of biopolitics: Lectures at the college de France 178–79.* Basingstoke, Palgrave Macmillan.

Foye, C., Clapham, D., & Gabrieli, T. 2017. Home-ownership as a social norm and positional good: Subjective wellbeing evidence from panel data. *Urban Studies*, 1–21. DOI: 10.1177/0042098017695478.

Freeden, M. 1998. Ideologies and political theory: A conceptual approach. *History and Theory*, 37, 139–140.

Galster, G. 2012. Neighbourhoods and their role in creating and changing housing. *In:* Clapham, D., Clark, W. & Gibb, K. (eds.) *Sage handbook of housing studies.* London, Sage.

George, V. & Wilding, P. 1976. *Ideology and social welfare.* London, Routledge and Kegan Paul.

Gibson, J. 1986. *The ecological approach to visual perception.* Hillsdale, NJ, Erlbaum.

Giddens, A. 1984. *The constitution of society: Outline of the theory of structuration.* Cambridge, Polity Press.

Gieryn, T. F. 2002. What buildings do. *Theory and Society*, 31, 35–74.

Gough, I. 2017. *Heat, greed and human need: Climate change, capitalism and sustainable wellbeing.* Cheltenham, Edward Elgar.

Grander, M. 2017. New public housing: A selective model disguised as universal? Implications of the market adaptation of Swedish public housing. *International Journal of Housing Policy*, 17(3), 335–352.

Granovetter, M. 1973. The strength of weak ties. *American Journal of Sociology*, 78, 1360–1380.

Gurney, C. M. 1999. Pride and prejudice: Discourses of normalisation in public and private accounts of home ownership. *Housing Studies*, 14, 163–183.

Haffner, M. E. A., Ong, R., Smith, S. J. & Wood, G. A. 2017. The edges of home ownership–the borders of sustainability. *International Journal of Housing Policy*, 17, 169–176.

Hagbert, P. 2016. "It's just a matter of adjustment": Residents' perceptions and the potential for low-impact home practices. *Housing, Theory and Society*, 33, 288–304.

Halpern, D. 1995. *Mental health and the built environment: More than bricks and mortar?* Abingdon, OX, Taylor & Francis.

Harmer, J. 2009. *Pension review report.* Canberra, Australian Government.

Haughwout, A., Peach, R., Sporn, J. & Tracy, J. 2012. *The supply side of the housing boom and bust of the 2000s.* New York, Federal Reserve Bank.

Hayek, F. A. V. 1944. *The road to serfdom.* London, Routledge.

Healey, P. 1992. An institutional model of the development process. *Journal of Property Research*, 9, 33–44.

Healey, P. 1997. *Collaborative planning: Shaping places in fragmented societies.* Basingstoke, Macmillan.

Hedman, E. 2008. *A history of the Swedish system of non-profit municipal housing.* Stockholm, Boverket.

Henning, C. & Lieberg, M. 1996. Strong ties or weak ties? Neighbourhood networks in a new perspective. *Scandinavian Housing and Planning Research*, 13, 3–26.

Henry, K., Harmer, J., Piggott, J., Ridout, H. & Smith, G. 2009. *Australia's future tax system: Report to the treasurer.* Canberra, Australian Government.

Hills, J. 2012. Getting the measure of fuel poverty: Final report of the fuel poverty review. CASE Report No 72. London, Centre for the Analysis of Social Exclusion.

Hincks, S. & Robson, B. 2010. *Regenerating communities first neighbourhoods in Wales.* York, Joseph Rowntree Foundation.

Hirsch, F. 1977. *Social limits to growth.* London, Routledge and Kegan Paul.

Hoekstra, J. 2003. Housing and the welfare state in the Netherlands: An application of Esping-Andersen's typology. *Housing, Theory and Society*, 20, 58–71.

Hogwood, B. W. & Gunn, L. A. 1984. *Policy analysis for the real world.* Oxford, Oxford University Press.

Holme, A. 1985. *Housing and young families in East London.* London, Routledge and Kegan Paul.

Holterman, S. 1975. Areas of urban deprivation in Great Britain: An analysis of 1971 census data. *In: Social trends.* London, HMSO.

Homelessness Directorate. 2003. *Reducing B & B use and tackling homelessness: What's working? A good practice handbook.* London, Homelessness Directorate.

Hulse, K. & Burke, T. 2016. Private rental housing in Australia: Political inertia and market change. *In:* Dufty-Jones, R. & Rogers, D. (eds.) *Housing in 21st century Australia: People, practices and policies.* London: Routledge.

Hulse, K., Jacobs, K., Arthurson, K. & Spinney, A. 2011. *At home and in place? The role of housing in social inclusion.* Melbourne, AHURI.

Hulse, K., Reynolds, M. & Yates, J. 2014. *Changes in the supply of affordable housing in the private rental sector for lower income households.* Melbourne, AHURI.

Jacobs, K., Atkinson, R.A.S., Colic Peisker, V. Berry, M. & Dalton, T. 2010. *What future for public housing: A critical analysis.* Melbourne, AHURI.

Jacobs, K., Kemeny, J. & Manzi, T. 1999. The struggle to define homelessness: A social constructionist approach. *In:* Hutson, S. & Clapham, D. (eds.) *Homelessness: Public policies and private troubles.* London, Cassell.

Johnsen, S. & Teixeira, L. 2010. *Staircases, elevators and cycles of change: Housing first and other housing models of support for homeless people with complex support needs.* London, Crisis.

Johnson, G. & Chamberlain, C. 2013. *Evaluation of the Melbourne street to home programme: 12 month outcomes.* Canberra, Department of Families, Housing, Community Services and Indigenous Affairs, HomeGround Services and The Salvation Army.

Johnson, G., Parkinson, S. & Parsell, C. 2012. Policy shift or programme drift? Implementing Housing First in Australia. AHURI Final Report no. 184 Melbourne.

Johnston, R., Poulsen, M. & Forrest, J. 2007. The geography of ethnic residential segregation: A comparative study of five countries. *Annals of the association of American Geographers,* 97, 713–738.

Jones, P. 2012. Housing: From low energy to zero carbon. *In:* Clapham, D., Clark, W. & Gibb, K. (eds.) *Sage handbook of housing studies.* London, Sage.

Jones, P. & Evans, J. 2008. *Urban regeneration in the UK.* London, Sage.

Jupp, B. 1999. *Living together: Community life on mixed tenure estates.* London, Demos.

Karvonen, A. 2013. Towards systemic domestic retrofit: A social practices approach. *Building Research & Information,* 41, 563–574.

Kemeny, J. 1981. *The myth of home-ownership: Private versus public choices in housing tenure.* London, Routledge and Kegan Paul.

Kemeny, J. 1991. *Housing and social theory.* London, Routledge.

Kemeny, J. 1993. The significance of Swedish rental policy: Cost renting: Command economy versus the social market in comparative perspective. *Housing Studies,* 8, 3–15.

Kemeny, J. & Lowe, S. 1998. Schools of comparative housing research: From convergence to divergence. *Housing Studies,* 13, 161–176.

King, P. 2009. Using theory or making theory: Can there be theories of housing? *Housing, Theory and Society,* 26, 41–52.

Klein, N. 2014. *This changes everything: Capitalism versus the climate.* New York, Simon and Schuster Paperbacks.

Knoll, K. A., Schularick, M. A. & Steger, T. A. 2017 No price like home: Global house prices, 1870–2012. *American Economic Review,* 107(2), 331–353.

Knox, P. 2008. *Metroburbia USA.* New Brunswick, NJ, Rutgers University Press.

Larimer, M. E., Malone, D. K., Garner, M. D., Atkins, D. C., Burlingham, B., Lonczak, H. S., Tanzer, K., Ginzler, J., Clifasefi, S. L., Hobson, W. G. & Marlatt, G. A. 2009. Health care and public service use and costs before and after provision of housing for chronically homeless persons with severe alcohol problems. *Jama-Journal of the American Medical Association,* 301, 1349–1357.

Lawson, J. 2006. *Critical realism and housing studies.* London, Routledge.

Lee, C. L. & Reed, R. G. 2014. The relationship between housing market intervention for first-time buyers and house price volatility. *Housing Studies,* 29, 1073–1095.

Lee, Y. & Ku, Y. 2007. East Asian welfare regimes: Testing the hypothesis of the developmental welfare state. *Social Policy & Administration,* 41, 197–212.

Leibfried, S. 1992. Towards a European welfare state? On integrating poverty regimes into the European Community. *In:* Ferge, Z. & Kolberg, J. (eds.) *Social policy in a changing Europe.* Frankfurt/Main, Campus.

Leishman, C. & Rowley, S. 2012. Affordable housing. *In:* Clapham, D., Clark, W. & Gibb, K. (eds.) *Sage handbook of housing studies.* London, Sage.

Lewis, O. 1966. *La Vida: A Puerto Rican family in the culture of poverty.* York, Random House.

Lindblom, C. 1965. *The intelligence of democracy.* New York, Free Press.

Lindbom, A. 2001. Dismantling Swedish housing policy. *Governance,* 14, 503–526.

Louise, A. R. & Houston, D. (2012) Low carbon housing: A 'green' wolf in sheep's clothing? *Housing Studies,* 28(1), 1–9.

Lukes, S. 1974. *Power: A radical view.* Basingstoke, Macmillan.

Lundqvist, L. J. 1987. Sweden's housing policy and the quest for tenure neutrality. *Scandinavian Housing and Planning Research,* 4, 119–133.

Maclennan, D. 1982. *Housing economics: An applied approach.* London, Longman.
Maclennan, D. & Miao, J. 2017. Housing and capital in the 21st century. *Housing, Theory and Society,* 34, 127–145.
Madden, D. J. A. & Marcuse, P. A. 2016. *In defense of housing: The politics of crisis.* London, Verso.
Maller, C., Horne, R. & Dalton, T. 2012. Green renovations: Intersections of daily routines, housing aspirations and narratives of environmental sustainability. *Housing, Theory and Society,* 29, 255–275.
Malpass, P. 2005. *Housing and the welfare state: The development of housing policy in Britain.* Basingstoke, Palgrave Macmillan.
Malpass, P. 2011. Path dependence and the measurement of change in housing policy. *Housing, Theory and Society,* 28, 305–319.
Malpezzi, S. & Maclennan, D. 2001. The long-run price elasticity of supply of new residential construction in the United States and the United Kingdom. *Journal of Housing Economics,* 10, 278–306.
Mandic, S. & Clapham, D. 1996. The meaning of home ownership in the transition from socialism: The example of Slovenia. *Urban Studies,* 33(1), 83–97.
Marsh, A., Gordon, D., Pantazis, C. & Heslop, P. 1999. *Home sweet home: The impact of poor housing on health.* Bristol, Policy Press.
Matlack, J. L. & Vigdor, J. L. 2008. Do rising tides lift all prices? Income inequality and housing affordability. *Journal of Housing Economics,* 17, 212–224.
Mats, W., Roland, A. & Kerstin, K. 2011. Rent control and vacancies in Sweden. *International Journal of Housing Markets and Analysis,* 4, 105–129.
Matznetter, W. & Mundt, A. 2012. Housing and welfare regimes. *In:* Clapham, D., Clark, W. & Gibb, K. (eds.) *Sage handbook of housing studies.* London, Sage.
Maxton, G. & Randers, J. 2016. *Reinventing prosperity: Managing economic growth to reduce unemployment, inequality, and climate change.* Vancouver, Greystone Books.
Mcmanus, A., Gaterell, M. R. & Coates, L. E. 2010. The potential of the code for sustainable homes to deliver genuine "sustainable energy" in the UK social housing sector. *Energy Policy,* 38, 2013–2019.
Meen, G. 2016. *European house price clubs.* Reading, University of Reading.
Meen, G. et al. 2005. *Affordability targets: Implications for housing supply.* London, Office of Deputy Prime Minister.
Minton, A. 2017. *Big capital: Who is London for?* London, Penguin Books.
Mirrlees, J. & Adam, S. 2011. *Tax by design: The Mirrlees review,* vol. 2. Oxford, OUP.
Mowen, J. C. 1987. *Consumer behavior.* New York, Macmillan; London, Collier Macmillan.
Moy, C. 2012. Rainwater tank households: Water savers or water users? *Geographical Research,* 50, 204–216.
Muñoz, S. 2017. A look inside the struggle for housing in Buenos Aires, Argentina. *Urban Geography,* 38, 1252–1269.
Murie, A. 2016. *The right to buy: Selling off public and social housing.* Bristol, Policy Press.
Murray, C. & Clapham, D. 2015. Housing policies in Latin America: Overview of the four largest economies. *International Journal of Housing Policy,* 15, 347–364.
Musterd, S. 2012. Ethnic residential segregation: Reflections on concepts, levels and effects. *In:* Clapham, D., Clark, W. & Gibb, K. (eds.) *Sage handbook of housing studies.* London, Sage.
Musterd, S. & Ostendorf, W. (eds) 1998. *Urban segregation and the welfare state: inequality and exclusion in western cities.* London, Routledge.
Nasar, J. 1993. Connotative meanings of house styles. *In:* Arias, E. (ed.) *The meaning and use of home.* Aldershot, Avebury.
Neale, J. 1997. Homelessness and theory reconsidered. *Housing Studies,* 12, 47–61.
Nielsen, B. G. 2010. Is breaking up still hard to do? Policy retrenchment and housing policy change in a path dependent context. *Housing, Theory and Society,* 27, 241–257.

Niner, P. 1989. *Homelessness in nine local authorities: Case studies of policy and practice*, London, HMSO.
Nussbaum, M. C. 2011. *Creating capabilities: The human development approach*. Cambridge, MA; London, Belknap.
OECD 2008. *Growing unequal? Income distribution and poverty in OECD countries*. Paris, OECD.
OECD 2017. *How's life?*. Paris, OECD.
OFT. 2008. *Home building in the UK*. London, OFT.
Organo, V., Head, L. & Waitt, G. 2012. Who does the work in sustainable households? A time and gender analysis in New South Wales, Australia. *Gender, Place & Culture*, 20, 559–577.
Oxley, M. & Marietta, H. 2010. *Housing taxation and subsidies: International comparisons and the options for reform*. York, Joseph Rowntree Foundation.
Padgett, D. K. 2007. There's no place like (a) home: Ontological security among persons with serious mental illness in the United States. *Social Science & Medicine*, 64, 1925–1936.
Parker, G. & Doak, J. 2012. *Key concepts in planning*. London, Sage.
Pawson, H. 2007. Local authority homelessness prevention in England: Empowering consumers or denying rights? *Housing Studies*, 22, 867–883.
Pawson, H. & Gilmour, T. 2010. Transforming Australia's social housing: Pointers from the British stock transfer experience. *Urban Policy and Research*, 28, 241–260.
Pawson, H., Netto, G., Jones, C., Wager, F., Fancy, C. and Lomax, D. 2007. *Evaluating homelessness prevention*. London, Communities and Local Government.
Pebley, A. & Vaiana, A. 2002. *In our backyard*. Santa Monica, CA, Rand Corp.
Pérez-Lombard, L., Ortiz, J. & Pout, C. 2008. A review on buildings energy consumption information. *Energy and Buildings*, 40, 394–398.
Phillips, R., Parsell, C., Seage, N. & Memmott, P. 2011. *Assertive outreach AHURI position paper*. Brisbane, AHURI, Queensland Research Centre.
Piketty, T. 2014. *Capital in the twenty-first century*. Cambridge, MA, The Belknap Press of Harvard University Press.
Popkin, S., Katz, M., Cunningham, K., Brown, J., Gustafson, J. & Turner, M. 2004. *A decade of Hope VI: Research findings and policy challenges*. Washington, DC, Urban Institute.
Psilander, K. 2012. Managing production costs of small and large developers in Sweden: A case study on multi-family construction. *International Journal of Construction Education and Research*, 8, 47–62.
Ravetz, A. 2001. *Council housing and culture: The history of a social experiment*. London, Routledge.
Raworth, K. 2017. *Doughnut economics: Seven ways to think like a 21st century economist*. London, RH Business Books.
Reid, L. & Houston, D. 2012. Low carbon housing: a green wolf in sheep's clothing? *Housing Studies*, 28(1) 1–9.
Rolnik, R. 2013. Late neoliberalism: The financialization of homeownership and housing rights. *International Journal of Urban and Regional Research*, 37, 1058–1066.
Ronald, R. (ed.) 2008. *The ideology of home ownership: Homeowner societies and the role of housing*. London, Palgrave.
Røpke, I. 2009. Theories of practice: New inspiration for ecological economic studies on consumption. *Ecological Economics*, 68, 2490–2497.
Rugg, J. & Rhodes, D. 2008. *The private rental sector: Its contribution and potential*. York, Centre for Housing Policy.
Ruming, K. 2014. Social mix discourse and local resistance to social housing: The case of the nation building economic stimulus plan, Australia. *Urban Policy and Research*, 32, 163–183.
Ruming, K. 2016. Reviewing the social housing initiative: Unpacking opportunities and challenges for Community Housing provision in Australia. *In*: Dufty-Jones, R. & Rogers, D. (eds.) *Housing in 21st century Australia: People, practices and policies*. Abingdon, Routledge.

Ruming, K., Mee, K. & McGuirk, P. 2004. Questioning the rhetoric of social exclusion: Courteous community of hidden hostility. *Australian Geographical Studies* 42(2), 234–248.
Ruonavaara, H. 2005. How divergent housing institutions evolve: A comparison of Swedish tenant co-operatives and Finnish shareholders' housing companies. *Housing, Theory and Society*, 22, 213–236.
Ruonavaara, H. 2012. Home ownership and Nordic housing policies in retrenchment. *In:* Ronald, R. & Elsinga, M. (eds.) *Beyond home ownership: Housing, welfare and society*. London, Routledge.
Ryan-Collins, J., Lloyd, T. P., Macfarlane, L. & Muellbauer, J. 2017. *Rethinking the economics of land and housing*. London, Zed Books Ltd.
Rybczynski, W. 2007. *Last harvest: How a cornfield became New Daleville*. New York, Scribner.
Sahlin, I. 2005. The staircase of transition. *Innovation: The European Journal of Social Science Research*, 18, 115–136.
Samuel, F. 2018. *Promoting design value*. Glasgow, Cache Working Paper.
Sandel, M. J. 2013. *What money can't buy: The moral limits of markets*. London, Penguin.
Sassen, S. 1991. *The global city: New York, London, Tokyo*. Princeton, NJ, Princeton University Press.
Savage, M., Bagnall, G. & Longhurst, B. 2005. *Globalisation and belonging*. London, Sage.
Sawtell, M. 2002. *Lives on hold: Homeless families in temporary accommodation*. London. Maternity Alliance.
Sayer, A. 2015. *Why we can't afford the rich*. Bristol, Policy Press.
Scase, R. 1999. *Britain towards 2010: The changing business environment*. London, DTI.
Schwartz, A. 2015. *Housing policy in the United States*. New York, Routledge.
Schwartz, H. M. & Seabrooke, L. 2009. *The politics of housing booms and busts*. Basingstoke, Palgrave Macmillan.
Sen, A. 1985. *Commodities and capabilities*. Amsterdam; Oxford, OUP.
Sen, A. 2010. *The idea of justice*. London, Penguin.
Shelter NSW. 2011. *Housing Australia factsheet*. Sydney, Shelter NSW.
Shinn, M. 2007. International homelessness: Policy, socio-cultural, and individual perspectives. *Journal of Social Issues*, 63, 657–677.
Shove, E. 2003. Converging conventions of comfort, cleanliness and convenience. *Journal of Consumer Policy*, 26, 395–418.
Shove, E., Pantzar, M. & Watson, M. 2012. *The dynamics of social practice: Everyday life and how it changes*. London, Sage.
Simon, H. A. 1947. *Administrative behavior: A study of decision-making process in administrative organisation*. New York, Macmillan.
Somerville, C. T. 1999. The industrial organization of housing supply: Market activity, land supply and the size of homebuilder firms. *Real Estate Economics*, 27, 669–694.
Somerville, P. 1989. Home sweet home: A critical comment on Saunders and Williams. *Housing Studies*, 4, 113–118.
Somerville, P. 1992. Homelessness and the meaning of home: Rooflessness or rootlessness? *International Journal of Urban and Regional Research*, 16, 529–539.
Somerville, P. (ed.) 1999. *The making and unmaking of homelessness legislation*. London, Cassell.
Speak, S. 2004. Degrees of destitution: A typology of homelessness in developing countries. *Housing Studies*, 19, 465–482.
Srna, M. & David, C. 1996. The meaning of home ownership in the transition from socialism: The example of Slovenia. *Urban Studies*, 33, 83–97.
Standing, G. 2017. *The corruption of capitalism: Why rentiers thrive and work does not pay*. London, Biteback.
Steg, L. & Vlek, C. 2009. Encouraging pro-environmental behaviour: An integrative review and research agenda. *Journal of Environmental Psychology*, 29, 309–317.
Stephens, M. 2001. Building society demutualisation in the UK. *Housing Studies*, 16, 335–352.

Stephens, M. 2011. *Tackling housing market volatility*. York, Joseph Rowntree Foundation.

Stephens, M., Fitzpatrick, S., Elsinga, M., Steen, G. & Chzhen, Y. 2010. *Study on housing exclusion: Welfare policies, labour market and housing provision*. Brussels, European Commission.

Stephens, M. & Van Steen, G. 2011. "Housing poverty" and income poverty in England and the Netherlands. *Housing Studies*, 26, 1035–1057.

Stone, M. E. 2006. What is housing affordability? The case for the residual income approach. *Housing Policy Debate*, 17, 151–184.

Streeck, W. & Thelen, K. 2005. Introduction: Institutional change in advanced political economies. *In:* Streek, W. & Thelen, K. (eds.) *Beyond continuity: Institutional change in advanced political economies*. Oxford, Oxford University Press.

Strengers, Y. 2011. Negotiating everyday life: The role of energy and water consumption feedback. *Journal of Consumer Culture* 11(19), 319–338.

Sunikka-Blank, M. & Galvin, R. 2012. Introducing the prebound effect: The gap between performance and actual energy consumption. *Building Research & Information*, 40, 260–273.

Suttles, G. 1972. *The social construction of communities*. Chicago, Chicago University Press.

Taylor, E. & Dalton, T. 2016. Keynes in the antipodes: The housing industry, First Home Owner Grants and the global financial crisis. *In:* Dufty-Jones, R. & Rogers, D. (eds.) *Housing in 21st century Australia: People, practices and policies*. London, Routledge.

Thomas, A. & Niner, P. 1989. *Living in temporary accommodation: A survey of homeless people*. London. DOE.

Torgerson, U. 1987. Housing: The wobbly pillar under the welfare state. *In:* Turner, B. Kemeny, J. & Lundqvist, L. (eds.) *Between state and market: Housing in the post-industrial era*. Stockholm, Almqvist and Wiksell.

Toro, P. A. 2007. Toward an international understanding of homelessness. *Journal of Social Issues*, 63, 461–481.

Tsemberis, S. 2010 *Housing First: The pathways model to end homelessness for people with mental illness and addiction manual*. Minnesota, Hazelden.

Tsemberis, S., Gulcur, L. & Nakae, M. 2004. Housing first, consumer choice, and harm reduction for homeless individuals with a dual diagnosis. *American Journal of Public Health*, 94, 651–656.

Turner, B. 1997a. Housing cooperatives in Sweden: The effects of financial deregulation. *The Journal of Real Estate Finance and Economics*, 15, 193–217.

Turner, B. 1997b. Municipal housing companies in Sweden: On or off the market? *Housing Studies*, 12, 477–488.

Turner, B. 1999. Social housing finance in Sweden. *Urban Studies*, 36, 683–697.

Turner, B. & Whitehead, C. 2002. Reducing housing subsidy: Swedish housing policy in an international context. *Urban Studies*, 39, 201–217.

van Kempen, R. & Bolt, G. 2012. Social consequences of residential segregation and mixed neighbourhoods. *In:* Clapham, D., Clark, W. & Gibb, K. (eds.) *Sage handbook of housing studies*. London, Sage.

Wacquant, L. 2008. *Urban outcasts*. Cambridge, Polity Press

Wacquant, L. C. J. D. 2009. *Punishing the poor: The neoliberal government of social insecurity*. Durham, NC, Duke University Press.

Weesep, J. V. 1987. The creation of a new housing sector: Condominiums in the United States. *Housing Studies*, 2, 122–133.

Wilcox, S., Perry, J. & Williams, P. 2016. *UK housing review*. Coventry, CIH.

Wilkinson, R. & Pickett, K. 2010. *The spirit level: Why equality is better for everyone*. London, Penguin Books.

Willmott, P. 1986. *Social networks informal care and public policy*. London, Policy Studies Institute.

Wissoker, P. 2016. Putting the supplier in housing supply: An overview of the growth and concentration of large homebuilders in the United States (1990–2007). *Housing Policy Debate*, 26, 536–562.
Woodhall-Melnik, J. R. & Dunn, J. R. 2016. A systematic review of outcomes associated with participation in Housing First programs. *Housing Studies*, 31, 287–304.
World Commission on Environmental Development. 1987. *Our common future*. Oxford, OUP.
Yates, J. 2012. Housing subsidies. *In:* Clapham, D., Clark, W. & Gibb, K. (eds.) *Sage handbook of housing studies*. London, Sage.
Yates, J. & Milligan, V. 2007. *Housing affordability: A 21st century problem*. Melbourne, AHURI.
Yolande, S. 2011. Negotiating everyday life: The role of energy and water consumption feedback. *Journal of Consumer Culture*, 11, 319–338.
Young, M. & Willmott, P. 1957. *Family and kinship in east London*. London, Routledge and Kegan Paul.
Zhou, J. & Ronald, R. 2017. The resurgence of public housing provision in China: The Chongqing programme. *Housing Studies*, 32, 428–448.

Index

Note: Italicized page numbers indicate a figure on the corresponding page. Page numbers in bold indicate a table on the corresponding page.

above-minimum incomes 158
accountability 13, 197–198
Acrosanti 186
Activities of Daily Living (ADL) 66–67
affordability of housing: demand-side subsidies and 138; government intervention in 126; housing policy and 21; income inequality and 122–124; overview 18, 65, 66, 73, 75, 199; public sector tenants 147
Agencia de Administración de Bienes del Estado (AABE) 55
alternative development movement 186
Anthropocene era 178
anti-union legislation 3
architectural distinctiveness 68
Argentina, homelessness 167–168
Argentina, housing market 53–55
asbestos 71
asset-based welfare 27
Australia: eco-cities in 185–186; First Home Ownership Grants 131–133; Housing First 174–175; rent assistance 129–130
average income levels 35

balanced communities 96
Bank of England 59
barriers to housing entry 60
basic capabilities 21
'basket of goods' comparison 123
Bed & Breakfast (B&B) hotels 165
behavioural economics 42
benefits policies 126
Berber house 66
Boarding Houses Act (2012) 77
bostadsrätt sector in Sweden 49, 49–50
British Building Societies 56, 58–60
British Housing Associations 12, 143, 150
British public housing case study 67–68

capital appreciation 157–158
capitalism: as corporatist market 32; income/wealth inequality and 121; rentier capitalism 46; residential capitalism 31–33, 35–36; 'varieties of capitalism' approach 33
carbon dioxide emissions 180
carbon-neutral housing 74, 178, 183–184
carbon tax 191
casas tomadas, defined 167
caveat emptor (buyer beware) 72
charitable organisations 41
China: homelessness in 160–161; housing regimes 38; housing shortages 117; public housing in 146–147; state-provided housing 151
Clark, John Bates 43
classical political economy 43
class status of neighborhoods 84
climate change 178, 179–182
coalition-building strategies 11, 17
commodification 3, 6–7, 32, 38, 194
Commonwealth Additional Grant (CAG) 132
Commonwealth-State Housing Agreement (CSHA) 144
communism 27, 152
Communities First programme in Wales 97–99
community and social interaction 85–87
community housing 145
competitive advantage 61
concomitant freedom 3
conditionality of welfare benefits 17
condominium housing markets 47–49, 60
Consumer Protection Act (2010) 57
'convergence' argument 31
conversion factors 21, 30
co-ordinate infrastructure provision 148
corporatist market 32, 134
cosmopolitanism 87

cost-covering rents 134
council housing 68
credit controls 60
crisis homelessness 160
critical junctures in path dependence 28
critical policy analysis 17
cultural mixing 89
culture of poverty 90, 97

Danish housing policy 30
decommodification 25, 34
demand assessment in housing supply 106–108
demand-side subsidies 127, 128, 137–139
demand subsidies 116
demographics of neighborhoods 84
dependency culture 92
direct provision 13, 196, 205
discourse 16–17, 198
discrimination in housing market 62
discrimination in lending practices 52–53
disequilibrium 45
displacement, defined 30
distribution of housing: government intervention 125–140; housing and wealth 124–125; income and wealth 118–120, **119**; introduction to 18, 118; social justice and 118, 120–124; summary of 140–141; types of intervention 127–137
Dodd-Frank Wall Street Reform 57
domestic practices 67
drift, defined 30
dual rented sectors 27
dysfunctional communities 112

Eastern European housing regime 6
eco-cities in Australia 185–186
Ecocity Builders 186
economic recession 104
efficiency mechanism of path dependence 28
elasticity 18
elective belonging 86
energy conservation 182
English language 65, 110
environmental sustainability: climate change 178, 179–182; evaluation of 186–191; government intervention 70, 74, 182; housing policy and 182–186; introduction to 2, 178–179; low-energy housing 183–184; in neighborhoods 185–186; retrofitting houses 184–185; summary of 191; in Sweden 189–190
environments of neighborhoods 84
equality in housing 118
escalator areas, defined 97
ethnic minorities 3–4, 51, 87, 91
European Commission 167

European Typology on Homelessness and Social Exclusion (ETHOS) 160
event-sequence model 102
exhaustion, defined 30

fairness in housing 118
family-owned land 27
Family Planning Certificate 161
Fannie Mae 56
Federal Housing Agency (FHA) 48
finance provision 13
financialisation 3, 35, 55, 101–102, 111, 121, 195, 204
fire hazards 78, 79–80
First Home Ownership Grants (FHOGs) 131–133
fixed rate mortgages 53
floating populations 160–161
FONAVI (Fondo Nacional de la Vivienda, National Housing Fund) 54
foreign ownership of housing 109
Foucauldian perspective 16
freedoms to achieve well-being 21
Friedman, Milton 3

gas leaks 71
generational divides in housing 120–121
gentrification 83, 91, 97
geological conditions 103
Gini coefficient 118–119, **119**
Global Financial Crisis (GFC) 3, 31–33, 35, 56–57, 121, 194
globalisation 121
global warming 179–180
good house 65, 66–70
government intervention: Australian rent assistance 129–130; environmental sustainability 182; First Home Ownership Grants 131–133; homelessness 171–172; housing distribution 125–140; housing market 46, 62; housing subsidies in Sweden 135–137; housing supply 105–108; impact of 138–140; objectives in homelessness 169–170; public housing in China 146–147; quality of housing 70–76; reasons for 91–93; rental housing assistance 130–131; rent-setting procedures in Sweden 133–135; tools and strategies 112–116; *see also* state-provided housing
government macro-economic policies 121–122
Grenfell Tower in the UK 79–80
gross domestic product (GDP) 54, 118
guidance intervention 13, 197, 206
Guidelines for Housing Policy (*Riktlinjer för bostadspolitiken*) 135

Harmer Pension Review 129–130
harm to dwellers 70–73
Henry Tax Review 130
Heygate estate, London case study 91–92
high-income housing market 104
high-risk mortgages 56–57
home *see* house and home
Home Deposit Assistance Grant 132
homelessness: in Argentina 167–168; in China 160–161; defined 160; discourses of 159–169; government intervention objectives 169–170; housing regime features 170–171; housing rights 171; introduction to 159; neoliberalism and 200; personal support 172; policy evaluations 175–176; prevention efforts 172–175; shelter provisions 171–172; summary 176–177; tools and mechanisms for 170–172; in UK 162–166, *165*
Homelessness Act (2002) 164
homeowner housing satisfaction 63
Hope VI programme 99–100, 155
house and home: affordances of home 65, 66; good house 65, 66–70; government intervention in quality of housing 70–76; harm to dwellers 70–73; housing quality 80–81; introduction to 65–66; market failure 74–75; policy tools 76–80; rooming houses 77–79; societal interest in 73–74; summary of 82
housebuilding industry in US 109–111
household budgets 126
houseless, defined 160
house price appreciation 62, 199
Houses in Multiple Occupation (HMOs) 77
Housing (Homeless Persons) Act (1977) 162, 163
Housing Action Areas 96
Housing Choice Vouchers 130–131
housing commodity 1
housing finance 32
Housing Finance Network 54
Housing First 172–175
housing market: Argentina 53–55; British Building Societies 58–60; constructing tenure 46–51; evaluation of 61–63; GFC case study 56–57; introduction to 41–42; as market commodity 44–46; market failure 74–75; nature of 43–47; policy tools 60–61; regulation of lending 52–58; summary of 64; tenant-owned apartment sector *49*, 49–50; transactions 52
housing outcomes 36–37
housing pathways 18

housing policy: defined 11–14; discursive approach to 16–17; evaluation of 17–23; introduction to 1–7, 11; making of 14–17; overview of 8–10; political approach to 15–16; rational approach to 14–15; structural approach to 16; summary of 23
housing quality 80–81
housing regimes: homelessness and 170–171; housing outcomes 33–34; introduction to 24–25; path dependence 28–31; residential capitalism 31–33, 35–36; six examples 37–39, **39**; summary of 39–40; welfare regimes 25–28
housing rights for homeless 171
housing shortages 75, 127, 199–200
housing subsidies in Sweden 135–137
housing supply: government tools and strategies 112–116; housebuilding industry in US 109–111; introduction to 101–102; need and demand assessment 106–108; reasons for government intervention 105–108; reasons for low supply 108–112; summary of 116–117; supply systems 102–105
Housing White Paper 112, 113

ideology of home ownership 135
imbalanced neighborhoods 96
immigration and housing supply 106
improvement grants to owner-occupiers 91
incentives for effective regulation 78–79
income distribution of housing 118–120, **119**
income/wealth inequality 121, 200
Index of Multiple Deprivation 97
India, housing market 51
individualisation 3, 195
information intervention 13, 197, 206
infrastructure of neighborhoods 84
inner-city households 68
institutional strengthening 95
institutional structures 36
inter-bank lending 31
interest-only mortgages 53
International Monetary Fund (IMF) 12, 121
intervention in housing market 52
isolate areas, defined 97

Kennedy, John F. 48
Keynesian economic management 59

land lots 55
language games 11, 17
Latin American Federation of Banks (FELABAN) 54
Latin American housing regime 38

layering, defined 30
legal regularisation 95
legitimacy mechanism of path dependence 28
Lehmann Brothers 57
lending regulations in housing market 52–58
lifetime homelessness 166
Lindbom, Anders 136
literal homelessness 159–160
loan-to-value (LTV) ratios 56
Local Housing Allowance 153
location matters 86
low-energy housing 183–184
lowest-income groups 121, 150–151
low housing supply 108–112
low-income housing 138, 142

macro-economic impact of housing finance 32
Manson, Fred 92
market failure 74–75
marketisation 3, 194; *see also* housing market
market-oriented organisations 61
maximalist discourse on homelessness 161–162
mental health factors 71
middle-class neighbourhoods 131
migrant labourers 161
Mill, John Stuart 43
minimalist discourse on homelessness 161
minimal standards in housing 76
minimum income levels 35
minority segregation 87–88
Minton, Anita 92
Mirrlees Committee (UK) 138
mixed-scanning approach 14
mortgage repayment 53, 57, 124, 162
mortgage securitisation 32
multi-dwelling buildings 46–47
municipal housing companies (MHC) 143, 150, 153–154, 168

National Administration of Social Security (ANSES) 55
National Affordable Housing Agreement (NAHA) 129, 144–145
National Housing Strategy 130
National Housing Supply Council 132
need assessment in housing supply 106–108
neighborhoods: Communities First programme in Wales 97–99; defined 84; dynamics 87–88; environmental sustainability in 185–186; evaluating policies 99; Hope VI programme 99–100; introduction to 83; overview of 84–88; physical regeneration of 94–96; policy mechanisms 93–100; problems with 88–90; reasons for government intervention 91–93; Rosario Habitat programme, Argentina 94–96; segregation 94, 96–97; social exclusion 90, 94, 97; social interaction and community 85–87; summary of 100
neo-classical economics 61, 69–70, 105, 188
neoliberalism (neoliberal ideology): accountability 13, 197–198; alternative regime to 203–207; commodification 3, 6–7, 32, 38, 194; concepts 193–195; direct provision 13, 196, 205; discourse 16–17, 198; environmental sustainability 200–201; evaluation of outcomes 18; gentrification 91; homelessness 200; house price appreciation 62, 199; house price inflation 189; housing market 46, 61; housing regimes 37–38, 62; housing shortages 75, 127, 199–200; housing supply 111; income/wealth inequality 121, 200; individualisation 3, 195; information or guidance 13, 197, 206; introduction to 1–6, 192–193; marketisation 194; non-intervention 13, 198, 206; outcomes of 198–202; policies and practices 195–198; privatisation 47, 151–155, 193–194; prospects for change 202–203; regulation 12, 76, 195–196, 204; residential capitalism 4, 31–33, 35–36, 202; social well-being 201; subsidies 13, 127–128, 196–197; taxation 13, 103, 126, 127, 135, 138, 196–197, 205–206; volatility in house price 62, 199; welfare regime 27; *see also* affordability; financialisation
new-built housing 75
'New Deal' legislation 156
New Public Management 3
nimbyism, defined 115
Non-Governmental Organisations (NGOs) 12
non-intervention 13, 198, 206
non-profit rents 134
Nordic housing regimes 29

OECD 121
Office for National Statistics (ONS) 143
Office of Fair Trading (OFT) 114
'off the peg' housing designs 105
open auction system 52
opportunity housing 145
overcrowding concerns 72, 76
owner-occupation: affordability concerns 122–123; demand-side subsidy 131; equality in housing 120; improvement grants to 91; overview of 26–27, 41; political constituency concerns 158; 'pure' owner-occupancy sector 136; tenant-owned apartment 50

Pareto criterion 20
paternalism 21
path dependency in housing regimes 4, 28–31
'Peak Oil' concerns 180
'pepper-potted' houses 157
perimeter concept 2
peripheral concept 2
person-centred approach to homelessness 173
physical regeneration of neighborhoods 94–96
planning regulations 108
pluralistic bargaining 15–16
polarisation of housing circumstances 3
policy tools 60–61, 76–80, 93–100
political approach to housing policy 15–16
political characteristics of neighborhoods 85, 89–90
political-economic policy problems 2
political ideologies 6, 25, 34
post-Fordist period 32
poverty concerns 89, 121, 126
power imbalances 15
power mechanism of path dependence 28–29
price regulation system 27
private rental markets 52
private *vs.* public space 83
privatisation 3, 47, 151–155, 193–194
problem housing estates 68
productivist regime 38
Programa de Atención para Familias en Situación de Calle 168
Programa de Crédito del Bicentenario para la Vivienda Única Familiar (Pro.Cre.Ar) 54–55
Programas Federales (PF) 54
proximity characteristics of neighborhoods 85
psychosocial environment 86
Public Health and Wellbeing Act (2008) 77
public housing 146–147, 155–156, 158
public-private collaboration agreement 148
public sector tenants 147
public service characteristics of neighborhoods 84
'pure' owner-occupancy sector 136

quality of housing 70–76

racial mixing 89
rational approach to housing policy 14–15
redevelopment policies 94
Register, Richard 186
regulation 12, 76, 195–196, 204
renewal policies for neighborhoods 96
rental housing assistance 130–131

rental market: demand-side subsidies in 139; dual rented sectors 27; non-profit rents 134; overview of 51; 'soft' rent control system 134; state-owned apartments 47; state-provided housing 148–150; unitary rental market 27, 133; utility-based rents 133
Rent Assistance (RA) 129–130
rent 'comparability' principle *(hyresförhandlingslagens likhetsprincip)* 134–135
renter housing satisfaction 63
rentier economy 46, 101, 119
rent payments 124, 128
rent-setting procedures in Sweden 133–135
residential capitalism 4, 31–33, 35–36, 202
residential development process 101–102
residential social practices 65, 66
Residential Tenancies Act (1997) 77
retrofitting houses 184–185
Right-to-Buy policy 152, 155
risk prevention programmes 95
risk to residents in rooming houses 78
Ronald, Richard 135
roofless, defined 160, 162
rooming houses 77–79
Rosario Habitat programme, Argentina 94–96
rough sleeping 160, 171
rural idyll 67

SAP rating of houses 188
Scandinavian-style housing co-operatives 24
Scottish Community-Based Housing Association 47
segregation 87–88, 94, 96–97
self-esteem 34, 68, 71, 86
self-provision of housing 104–105
separateness concerns 88
settlement upgrading programme 94–96
severe weather events 181
shanty towns 160
shared-ownership housing 145
shopping online 86
single-family dwellings 136
skill-based technological changes 121
sleeping rough 160, 171
slum clearance programmes 83, 94
social constructionist perspective 16
social-democratic regimes 62, 102, 157
social-democratic welfare states 25, 26–27, 32
social exclusion 90, 94, 97
social housing 142, 144–145
social interaction and community 85–87
social interaction of neighborhoods 85
social justice 118, 120–124
social psychological factors 71
social solidarity 91

social welfare 126
social well-being 201
societal interest in home 73–74
'soft' rent control system 134
Soleri, Paolo 186
Southern European rudimentary welfare states 27
spatial segregation 68
'staircase' model 172–175
state-dominated supply process 105
state-owned apartments 47
state-provided housing: defined 143–144; development of 147–148; introduction to 142; management of 148–150; municipal housing companies in Sweden 153–154; objectives and roles 144–147; private and public rental 150–151; privatisation of 151–155; pros and cons 155–158; social housing 142, 144–145
state's pension fund (Fondo de Garantía de Sustentabilidad) 55
statist-developmentalist 32
statutory homelessness system 162–165
'step' rule *(trappningsregel)* 135
Stockholm Resilience Centre 179
stratification of housing 34, 38
structural approach to housing policy 16
subjective well-being 20
sub-prime loans 57
subsidies by government 13, 127–128, 196–197, 205–206
subsidy provision 13
'suck-it-and-see' manner 15
supplementation homelessness 160
supply-side subsidies 128, 137–139
supply systems in housing 102–105
supported housing 145
survival homelessness 160
sustainability *see* environmental sustainability
Sustainable Development Goals (SDGs) 179
SUSTAIN project 71
Sweden: *bostadsrätt* sector *49*, 49–50; environmental sustainability in 189–190; homelessness in 170–171; housing market 49; housing subsidies in 135–137; municipal housing companies in 153–154, 168; rent-setting procedures in 133–135; right to shelter 171
Swedish Housing Co-operatives 69
symbolically spoiled zones 90

taxation on housing 13, 103, 126, 127, 135, 138, 196–197, 205–206
Temporary Living Permit 161
tenancy contracts 145
tenant control 150

tenant-owned apartment sector *49*, 49–50
Tenant-Ownership Act (1930) 49
Tenant-Ownership Control Act (1942) 49, 50
Tenants Unions 150
tenure in housing market 46–51, 69
tenure neutrality 135–137
territorial indignity 90
Thatcher, Margaret 162
third-sector organisations 102
transaction costs 18, 45
transactions in housing market 52
transit areas, defined 97

unintentionally homeless 162
unitary rental market 27, 133
United Kingdom (UK): homelessness in 162–166, *165*; Right-to-Buy policy 152, 155; statutory homelessness system 162–165; welfare reform *165*, 165–166
United States (US): condominium market in US 47–49; housebuilding industry in 109–111; public housing 155–156; rental housing assistance 130–131
universal standards of space 72
urban planning 95
urban sprawl 115
use-value *(bruksvärde)* approach 133
utilitarian approaches 20
utility-based rents 133

value added tax (VAT) 127
value-free analysis 14
value gap 91
'varieties of capitalism' approach 31, 33
volatility in house price 62, 199
volume housebuilders (VHB) 112

Washington Consensus 38
wealth: capitalism and 121; housing and 124–125; income and 118–120, **119**; inequality 121, 200
wealth distribution of housing 118–120, **119**
welfare benefits 17
welfare ideologies 36
welfare reform *165*, 165–166
welfare regimes 25–28, 63
welfare retrenchment 3
well-being 69–70, 75, 81, 83, 85
Welsh Assembly Government 97
worker housing 145
working-class community 85
World Bank 12
World War I 67
World War II 144

zero-carbon housing 183